MORTAL STAKES

ALSO BY JAN E. DIZARD

Guns in America
(coeditor with Robert M. Muth and Stephen P. Andrews Jr.; 1999)

Going Wild:
Hunting, Animal Rights, and the Contested Meaning of Nature
(1994; revised and expanded 1999)

The Minimal Family
(coauthor with Howard Gadlin; 1990)

Social Change and the Family
(1968)

MORTAL STAKES

Hunters and Hunting in
Contemporary America

Jan E. Dizard

University of Massachusetts Press

Amherst and Boston

LC 2002010179
ISBN 1-55849-365-4 (library cloth); 384-0 (paper)

Designed by Milenda Nan Ok Lee
Set in Galliard
Printed and bound by The Maple-Vail Book Manufacturing Group

Library of Congress Cataloging-in-Publication Data

Dizard, Jan E., 1940–
    Mortal stakes : hunters and hunting in contemporary America / Jan E. Dizard
      p. cm.
Includes bibliographical references (p.) and index.
    ISBN 1-55849-365-4 (lib. cloth ed. : alk. paper)—
ISBN 1-55849-384-0 (pbk. : alk. paper)
    1. Hunting—Moral and ethical aspects—United States. I.
Title.
    SK14.3 .D59 2003
    179'.3—dc21
2002010179

British Library Cataloguing in Publication data are available.

But yield who will to their separation,
My object in living is to unite
My avocation and my vocation
As my two eyes make one in sight.
Only where love and need are one,
And the work is play for mortal stakes,
Is the deed ever really done
For Heaven and the future's stake.

—Robert Frost, "Two Tramps in Mud Time"

CONTENTS

ACKNOWLEDGMENTS

Heartfelt thanks must first be given to the men and women who submitted to my questions and so generously gave of their time. Without each of their perspectives and experiences, this book would not have been possible. Without their good humor and willingness to share, the book would have been much less fun to write.

Without a grant from Amherst College's H. Axel Schupf '57 Fund for Intellectual Life, it would have been vastly more difficult to conduct and transcribe the interviews. I am grateful for the College's and Axel Schupf's generous support.

A number of friends and colleagues have commented on early drafts or on conference presentations of one or another aspect of *Mortal Stakes*. John Organ and Kevin Richardson, as much as anyone else, spurred me to undertake systematic research of hunters. Shane Mahoney also gave me early encouragement as well as much to ponder. Mary Zeiss Stange's encouragement and sharp critiques have helped enormously. Bob Muth has been a steady source of interesting and helpful ideas. I met Jim Tantillo and Marc Boglioli just after I had completed my interviews and was in the early stages of making sense of all that I had heard. Marc had just finished his anthropological fieldwork with Vermont hunters, and comparing notes with him was enormously helpful. When his study is published, it will be a wonderful addition to our understanding of hunting and hunters. Jim's work on hunter ethics and justifications for hunting has also been helpful as I have tried to sort out the complexities that lie at the heart of hunting. It is friends and colleagues such as these who have made the writing of *Mortal Stakes* seem more like a stimulating conversation than work. I am grateful.

Clark Dougan has once again been a wonderfully supportive and patient editor. Indeed, the entire staff of the Press could not have been more helpful.

Two close friends, Jim Dowd and Cameron Cope, with whom I've hunted, among other things, for over thirty years, have been a constant source of delight. Were it not for the lessons they helped me learn about the meaning of the hunt as well as the meaning of friendship, this project would have been far shallower.

My wife, Robin, suffered cheerfully through innumerable interruptions to answer a question about syntax or to be a sounding board for one or another of my half-formed ideas. Her wide reading and attentive eye caught many interesting books and articles I no doubt would have remained ignorant of. For this, and for putting up with my passion for hunting, I am profoundly grateful.

Finally, I would like to record here a hope that many of the older hunters I interviewed also expressed: I hope that I will be able to share the intensities of the hunt with my grandson, George Thomas Dizard, and with my granddaughter, Nadia. It is to them that I dedicate *Mortal Stakes*.

MORTAL STAKES

PROLOGUE

A friend once remarked that all writing is autobiographical. Whether or not that is true for all writing, it is certainly true for this bit of writing. I did not come to the subject of this book by way of some dispassionate inquiry. On the contrary, I decided to write about hunters because, as an avid hunter myself, I was growing impatient with the ways both hunters and their critics talked about hunting. The rhetoric attacking as well as defending hunting had little correspondence with how the hunters I have known behaved and even less with how they thought. Long fascinated by the ways social and cultural change creates all sorts of eddies, cross-currents, and dead water, I was curious about how men and women who have embraced an activity that many view as anachronistic think about their society and culture. Have hunters somehow failed to keep up with the times? Are they stuck in the past, caught in an eddy that keeps them from adopting more refined pursuits? Are they just a bunch of good ol' boys who get their kicks by thumbing their noses at civilized people? Armed with my own experience, I set out to interview a small, randomly selected number of hunters so that I might draw a more accurate portrait, warts and all, of contemporary hunters. Since this is, in some sense, autobiographical, it makes sense to begin by sketching how I became a hunter.

I didn't start out a person destined to take up hunting. No one in my family hunted. My father, the only son of immigrant parents, grew up on a hardscrabble farm on the outskirts of Duluth, Minnesota. Though large and muscular, he was always more inclined to read than to enjoy the outdoors. Still, he somehow acquired a .22 rifle, as he tells it, when he was twelve or thirteen and promptly shot a rabbit. His shot was not a very good one, and he told me only once how he stood frozen with remorse as the rabbit frantically jumped and writhed and piteously shrieked in pain. He could not bring himself even to put the creature out of its misery. He fled the scene, the sights and sounds of the suffering he'd caused indelibly with him. My father hated violence of all kinds, and while he did not object to or try to persuade me not to hunt, he made it plain that he was not going to assist me in any way. I would have to wait until I was old enough to purchase my own gun and hunting license. My grandfather, with whom I fished often, was no help either, at least not until I was well along the way toward being able to hunt on my own. He was an observant though not orthodox Jew, and hunting was foreign to him—by definition, an animal that is killed by a bullet or an arrow cannot be kosher—and he'd had all he wanted of guns during his brief service in France in World War I.

I was also a frail kid and, like my father, bookish. Were I a child now, I'd likely be regarded a nerd by my schoolmates. It didn't help that I got polio when I was nine. Though I fully recovered, during late childhood and early adolescence I was awkward and slow—not quite the last one picked when sides were chosen for games but very nearly the last one. Ironically, though not surprisingly I suppose, the illness that made me a sorry physical specimen intensified my interest in the out-of-doors. Like a kid with his nose pressed against the candy store window, I devoured tales of trappers, buffalo hunters, and the sundry misfits and adventurers who matched wits with Indians and wildlife and managed to "tame a continent." The juxtaposition of their hardiness and my helplessness could not have been sharper.

My imagination was also nourished by the world in which I was growing up. One of my earliest recollections—I could have been no more than six—was looking out of the kitchen window of the duplex we were living in and seeing eight or so wolf pelts hanging from a line stretched along the garage wall. The man who lived in the other half of the house, a newspaper photographer, had had a very successful week trapping wolves in the north woods. The war had just ended and the prosperity that always seemed around the corner was still around the corner, at least in Duluth,

Minnesota. Even steadily employed men looked to hunting, trapping, and fishing as ways to ease pressure on tight family budgets. Hunting, trapping, and fishing were also fun—a chance to get away from the constraints of the household and "be with the boys" as well as an escape from the often dulling rhythms of factory work or the pressures of meeting deadlines.

My physical limitations did not affect my ability to fish, and my grand-father loved to fish. Fishing became my passion. It gave me particular pleas-ure to bring fish home, clean them, and then bask in the satisfaction of see-ing my contribution to the household economy eaten with relish. It made me feel somehow older and more competent than I was. But I wanted to hunt too. The wolf pelts, the deer draped over the rounded front fenders of a '49 Chevy or Ford, and the less tangible evidence of overheard con-versations of friends and neighbors talking of clouds of pheasants rising out of southern Minnesota or South Dakota cornfields or of a flight of geese coming into a shallow bay choked with wild rice made me ache for a chance to hunt.

By the time I was able to resume more or less normal physical activities, my grandfather had purchased a home beyond the city limits of Duluth—an early expression of sprawl—on roughly ten acres of mostly wooded land. He and my grandmother, both afflicted with heart disease that in those days dictated a sedentary lifestyle, became avid bird watchers and had the trees in front of most windows supporting one or more bird feeders. Of course, squirrels became their nemesis. When I was fourteen, my grand-father bought me a .22 rifle and offered me a bounty for each squirrel I dis-patched. Since I would gladly have paid for the privilege of shooting game, I was doubly overjoyed and commenced laying siege to squirrels. I got to be quite a good shot with that rifle but was conflicted about the squirrels. I'd learned from the hook and bullet magazines that I had been reading each month for years that hunters eat what they kill (varmints excepted). No one in my family was the least bit interested in eating squirrel. I skinned and cleaned the first couple I shot and, following a recipe I'd read in one of the magazines, cooked them up. I liked the result, but only my mother could be persuaded to taste a morsel. She found the flesh agreeable but ob-served that it would take a bushel of squirrels to feed what had become a statistically exemplary baby boom family.

I was thus in a bind. I could justify killing squirrels so that my grandpar-ents could better enjoy the birds that had brightened their lives, but I knew that this wasn't exactly what I had been yearning after. Hunting squirrels is challenging, but shooting squirrels at bird feeders is not. All I had to worry about was shooting in a safe direction and being sure not to shatter the feeders. These conditions were not difficult to meet. And then there was the meat. Though squirrels were abundant, I never killed enough of them at any one time to make so much as an appetizer much less a meal for a family of six (my grandparents would not touch the stuff), and I tired of going to all the bother of skinning and gutting one or two squirrels for my own dining pleasure. I had become a hired gun, not the hunter I'd dreamt of becoming.

My subsequent youthful forays into hunting were similarly freighted with difficulties, less of the ethical than of the practical sort. Surrounded by hunters and fascinated by the hunting stories I read, I nonetheless had no actual experience. A family friend who had grown up in a small town hidden in the north woods and whose family had relied heavily on the fruits of hunting gave me my first hands-on introduction to hunting and the woodcraft upon which hunting depends. Buddy Johnson showed me how to look for deer sign, how to pick out the places where ruffed grouse were most likely to be found, how to read a compass, what to do if lost, and a host of small little bits of lore and wisdom that I could have learned from books but which took on much more depth and significance coming from him, the voice of experience. Buddy died of a freak accident before I was out of college. We never got a chance to hunt together, but for what it is worth, in our own fashion we have hunted frequently and well together for many years now.

By the time I was old enough to have my own car and my own gun, my ardor for the hunt had waned—I was becoming a serious student, girls seemed more alluring than the autumn woods, and my car and girlfriend required cash that in turn obliged me to work on weekends. Still, I diverted some of my hard-earned money to the purchase of a cumbersome, bolt-action shotgun, a deer rifle, and a fancier .22 than the one my grandfather had given me. Friends and I would go out once in a while to shoot tin cans and hand-thrown clay pigeons, and I would occasionally try in vain to shoot a grouse on the wing. I was a decent rifle shot but the shotgun baffled me. I tried my hand at deer hunting only once back then. A friend and I went out opening day to a spot reputed to be thick with deer. We took up our prearranged posts, hunkered down against the cold, and waited for deer to make themselves known. It wasn't too long before my feet were numb, and the rest of me was craving motion. I remained still, just as the outdoor writers counsel, until I could stand it no longer. I stirred and began stamping my feet to get the circulation going. I tried my best to do this quietly so as not to scare off any deer, but I must have made just enough noise to sound like a deer. A shot rang out, and a bullet whizzed by, high and just to my left, sending bark flying from the tree against which I'd been leaning. Instinctively, I hit the ground. I tried to shout, but I was too terrified to utter so much as a bleat. There was one more shot, now comfortably well over my head. The second shot broke through my terror, and I let out a shout and fired two quick shots into the air. I lay motionless for what seemed an eternity before I carefully rose to my feet. My friend then appeared, expecting to see me with a deer after hearing those four shots. Neither of us had seen anyone nor even heard another hunter. Someone must have stealthily been on the move, and when he heard me scuffling he shot at what he presumed was a deer. When he realized his mistake he fled.

Any thoughts of going deer hunting again vanished when, only days after my scare, two friends were seriously wounded, one shot by a hunter who mistook him for a deer and the other by a freak accidental discharge of his own rifle. I promptly sold my deer rifle and used the proceeds to get a new autoloading shotgun. I lamely figured that what I lacked in skill with a shotgun I could make up with firepower. With five quick shots, surely I'd be able to get one of those damned elusive grouse. The odds seemed dauntingly in my favor. The woods were filled with grouse in northern Minnesota, and I could fill the air with a veritable cloud of lead with my new shotgun. Something had to fall from the sky. I practiced with clay pigeons, but I now know that all I was doing was practicing bad technique. Needless to say, I did not make much of a dent in the grouse population around Duluth. In fact, try though I might, I managed to kill but one grouse in those years and that event was depressing. I was hunting along one of the many rivers that flowed into Lake Superior when a grouse flushed from a tangle between me and the river. My shot caught the grouse just as it cleared the tree tops and was veering toward the river. My elation quickly turned sour as the grouse tumbled into the river and was swept away by the fast current. The river was too deep to wade, and I was not a strong enough swimmer to give even a moment's thought to diving into the cold current, so I watched helplessly as "my" grouse floated away.

There was, in short, nothing in my early experiences with hunting that would incline me to become an avid hunter. In fact, when I was accepted to graduate school, I sold the few guns I had acquired: I needed the cash. The combined intensities of graduate work, marriage, a new baby, then my first job, all taking place in the midst of the Civil Rights movement and the antiwar movement, in both of which my wife and I were heavily involved, pushed hunting deep into my subconscious, so deep that my wife was shocked when, after six years of marriage, I indicated a bit sheepishly that I was thinking about buying a shotgun and a bird dog. My wife was brought up a Quaker, and guns were an anathema to her. Even now, having become thoroughly inured to my hunting, she does not like the idea of guns in the house. In deference to her sensibilities, as well as for safety reasons, the only indications of there being a hunter in our home are a few game bird prints on various walls, an extensive collection of books on hunting and related topics, and a bird dog or two under foot. Oh—and a goodly number of cookbooks devoted to the preparation of game. Luckily for me, my wife has found most game to her liking, though to be honest she may at first have simply liked the fact that I did the cooking when game was involved.

The turn back to hunting came abruptly and without warning. In the summer of 1969 we moved to Amherst, where I was to begin teaching at Amherst College in the fall. On one of our many trips to the local discount stores to get some of the myriad small things needed to set up a household, I decided to take a back road that paralleled the Connecticut River. I was scouting the river for fishing opportunities. All of a sudden, two cock pheasants sprinted across the road just ahead of our car and headed into the cornfield that paralleled the road. I slammed on the brakes and, rapt, watched those two beautiful birds disappear into the corn. It was as if someone had attached electrodes to me and sent a charge through my limbs. I knew then and there that I would give hunting another try.

I must say that I picked up where I'd left hunting some eight or nine years earlier—though I was now in a position to buy a better and more appropriate shotgun, all the bad habits and incompetence I had nourished years earlier had stayed with me. I hunted occasionally that first fall in Massachusetts with two colleagues from the college, Jack Pemberton and Jim Chalmers, and, despite their generous sharing of know-how and the excellence of their bird dogs, I was only able to scratch down one pheasant (which, to this day, I am sure was actually shot by my companion, who graciously insisted that it had been my shot, not his, that had felled the bird). Frustration piled on top of frustration, but I was determined to overcome my pitiable ineptness. What proved crucial to my perseverance was not strength of character and certainly not encouragement from my wife; what drove me, and still drives me, is a complete fascination with dogs bred to hunt birds.

I grew up with dogs, but they were pets—a collie and an Airedale while living with my parents, then a beagle and a foxhound when I was married. My wife and I would take the beagle and foxhound to the outskirts of town where they could chase rabbits to their hearts' content, but that was to give them exercise, not to hone their or my hunting prowess. I'd never hunted over a dog until that first fall in Amherst. I will never forget the first time I saw Jim Chalmers's German wirehaired pointer, Ziggy, course full tilt through a field and suddenly stop dead in her tracks, frozen into the classic point: head stretched out, body slightly crouched, one foreleg lifted as if in mid-stride. I'd seen pictures galore of such poses and had read others' descriptions of the intensity of that moment, but seeing a dog on point was far more exciting than anything I had imagined. The October chill that had numbed my fingers left in a rush of adrenalin as I approached Ziggy. When the cock pheasant finally could stand the suspense no longer and took flight, cackling indignantly as his wings lofted him above the corn stubble in which he had tried to conceal himself, I was so excited that I forgot to bring my gun to my shoulder. I was smitten.

A month later, my wife, pregnant with our second child, our six-year-old son, and I went hunting for a bird dog puppy. I had perused ads in the local papers as well as in the Boston and New York papers. For some reason, one ad struck me, and I called to make sure puppies were still available. There were two left. We dropped everything and headed for a small town in central Massachusetts. By evening we were back home with my first bird dog, Elijah, a not-quite-three-month-old German shorthaired pointer. Elijah and I taught each other how to hunt birds. That dog and I spent at least an hour a day, rain, snow, heat, cold, going over the basic obedience training and, by stages, the more demanding and refined points involved in finding birds and retrieving them. Often that first winter Elijah and I would founder through deep snow following the tracks of a pheasant, his stubby tail wagging furiously with excitement. I banged pots at his meal times to acclimate him to loud noises, working up from pots to my son's cap pistol, then to a blank pistol of the sort used to start races and finally to the real thing, a shotgun. We lived well away from the center of town and I went around to the few homes that constituted our neighborhood to explain what all the noise was about. No one objected though I suspect some of our neighbors felt sorry for my wife and son. Elijah and I became quite competent pheasant finders, and in the bargain, Elijah turned out to be a very classy pointer.

By the following summer it was clear that with Elijah's capable nose we would have no trouble finding pheasants when October rolled around. The question remained: Could I learn how to shoot well enough to complete the moral contract I had entered into with Elijah? With a new baby to marvel at and an older child to reassure, and with Elijah and me still unraveling the mysteries of pheasants and the subtleties of our collaboration, time was short. Of course, everyone knows "professors have the summer off." Sure. There was plenty of midnight oil burned in our home that summer so that I would have the time to begin learning how to be at least reasonably effective with a shotgun. I joined the local rod and gun club and started shooting skeet a couple of times a week. My ineptitude was no doubt the subject of much mirth among the mostly working-class locals who were the club regulars. I suppose my long hair and VW van festooned with antiwar slogans and "Free Angela Davis" bumper stickers (the sixties were not quite over in the early 1970s) didn't help my standing with these fellows. At least they could take comfort in knowing that if the "revolution" my bumper stickers were promoting should arise, they would be perfectly safe with me shooting at them.

In fact, I did begin to improve by fits and starts, largely because whatever else these far more experienced men may have thought of me, they understood the fire that had been kindled in me and they were eager to coach and encourage me. There were and still are many things those folks and I disagree about, and there are many ways in which our tastes and styles diverge, but I came to realize that in places like the rod and gun club, many of the distinctions that bedevil us so greatly, including race and gender and class, are set aside. Liking guns, bird dogs, and the hunt are more than sufficient to earn a place among such men and women as I have learned from. Like the men and women I interviewed, the people for whom the rod and gun club served as a social center as well as a shooting range were a varied lot: a couple of professors, lots of carpenters, electricians, contractors, and maintenance men (and women), for the most part people who are working class. Some bore the marks of hard living, and some were braggarts. There were occasional references to tumultuous marriages and various versions of misspent youth. But even the coarsest amongst them would quickly become considerate and supportive when one of their number was in difficulty—all sorts of informal exchanges of goods and services flowed freely among the regulars; and when there was an illness or even worse calamity, the informal networks linking the regulars would buzz with offers of assistance. Refined ears might be offended by the banter and advocates of temperance would have grist for their mill, but all in all, these folks, many of whom I have now known for over thirty years, do not provide a very big target for those who would criticize hunters and firearms enthusiasts for their violent and antisocial ways. Indeed, they have led more respectable lives and steered clear of more bad habits than some of our highest elected officials, past and present. But I digress.

By the opening of pheasant season, I had become a fair wing shot and Elijah had become quite good at finding pheasants. Opening day fell on one of my most intense teaching days of the week, but I was determined to make at least a brief appearance in the field, and for the first time since Elijah and I began working together almost a year ago, it was time to see if we could put it all together. My daily outings with Elijah gave me something of a head start on many other hunters—I knew where there were certain to be pheasants. Some Elijah and I had flushed so many times over the course of the summer and early fall that they almost felt part of the family. Time was short so I decided to open the season and a whole new phase of my life in a small cover—I had not yet gotten used to saying "covert"—just over a half mile from my house. Elijah seemed to pay no particular attention to the long leather cased object I put in the van next to his dog crate. After that brief morning's outing, however, whenever I cased my shotgun he would race to the back door and wait with rapidly dwindling patience to get going.

In my eagerness I left a good half hour sooner than I should have—it was still too dark to hunt. I paced in the cold as if willing it would make the sunrise hurry. By the time it was light, I was chilled and my cold hands fumbled with the latch on Elijah's crate. I got the door open finally and Elijah bounded out and headed out into the familiar field. I self-consciously slid my shotgun out of its case, forced my cold fingers to grasp two shells from my vest pocket, and headed after Elijah. It took all of ten minutes for Elijah to "lock up," one of the terms bird hunters use to describe a dog on a solid point. Judging by Elijah's intensity, the bird was holding tight. Pheasants are strong fliers but they prefer to sneak away from danger, and if they flush at all, they do so at a distance. When a bird begins to sneak

9

away, some dogs will remain locked up until their handler commands them to relocate; other dogs will slowly melt out of a solid point and begin a slow, deliberate creep in pursuit of the fleeing bird. Elijah was a creeper but this time he remained as motionless as a rock. As I closed the distance between him and me, I brought my gun to port arms and steeled my nerves for the flush. I was almost alongside Elijah when the cock pheasant clattered out of the waist-high swale about five yards in front of me. I raised my gun to my shoulder, trying to recall all the advice I'd been given as I swung the barrel following the bird's flight. I took a deep breath and pulled the trigger. The bird veered sharply to the right and then dropped. I'd just shot my first pheasant. I could scarcely believe what had just happened. I turned to Elijah, still on point at my side, and hugged him as if he were a long lost friend.

My ecstasy was short lived. An experienced bird hunter, I quickly learned, never takes his eyes off the spot where a bird, especially a pheasant, lands. Even mortally wounded birds often have enough life left in them to run a short distance and vanish beneath a brush pile or the thick mat of grasses that form the base of uncultivated fields. "My" bird was not mortally wounded. Even though a rank beginner, I realized that my shot had been off center and the bird dropped because I'd broken one of his wings. I did not realize the implications of this. Elijah and I began searching for the bird. I headed toward the general area where the bird had landed and exhorted Elijah to find it. We crisscrossed the area to no avail. Triumphalism rapidly turned to remorse. I was mad at myself for having shot poorly, I was disappointed in Elijah for not finding the bird, and as he and I went vainly back and forth over the same ground, I began to question the whole enterprise. I was so absorbed in this misbegotten enterprise that I forgot all about my 9 o'clock class until it was almost too late. Luckily, I didn't have far to drive.

Late in the day, I took Elijah back to the fateful spot for one last attempt. We did find a dead bird, which some other hunter's dog or a predator had relieved of most of its breast meat. I consoled myself that it was the remains of "my" bird and that at least it had not suffered long. Then, with some of the morning's burden lifted, what had happened that morning came into clearer focus. I had trained Elijah to find and point birds and I had trained him to retrieve a canvas retrieving dummy that I had impregnated with commercially prepared pheasant scent. I thought that the connection between these two activities was as obvious to Elijah as it was to me. Had the pheasant I wounded

10

turned into a canvas dummy smelling vaguely like a pheasant, Elijah in all likelihood would have been on it in a trice. But the retrieving of a bird was, from
his point of view, not yet part of the deal we'd struck.

This came to me in a flash, as I contemplated the ransacked carcass of
what had been, just a few hours earlier, a magnificent creature. I could turn
back. Elijah would be a wonderful pet and I had plenty going for me in my
workaday life, not to mention a happy marriage and two wonderful kids.
Why did I need to pursue this, whatever it was? Light was fading and it was
my turn to prepare dinner and I had to prepare for another early class the
following day. Something, I cannot to this day name exactly what its source
was, held me back. The best I can offer, after years of reflection, is that I
knew, deep down, that I had invited Elijah into a bargain that required me
to honor his willingness to accept the offer. He needed to complete his part
of the bargain, which meant learning to retrieve the birds I somehow managed to scratch down; and I needed to make his work easier by becoming
a better wing shot. It would, as it turned out, take me a good deal longer
to become a passable wing shot than it took to make Elijah into a first-class
retriever. Neither process was pretty.

Hesitantly, I waved the disheveled remains of the pheasant in front of
Elijah and then threw the carcass a few yards away. I then gave Elijah the
command, "fetch." He bounded toward the bird and, after a sidewise
glance in my direction, mouthed the bird and, ever so reluctantly, picked
it up. With the damned thing dangling limply from his most reluctant jaws,
he turned to face me, as if to ask, "Is this really what you expect of me?" I
repeated the command, "fetch," and he came toward me, head down to
convey his sense of the disagreeableness of all this. He delivered the sem-
blance of the bird, as he'd been trained to deliver the canvas dummy. I
praised him lavishly, having already decided to test both his and my resolve
further. I laid the pheasant carcass a couple of feet away and physically established Elijah on point, facing the remains of the pheasant. Then I picked
up the carcass and threw it as far as I could with my left hand and, while it
was still in midair, I fired my shotgun into the air with my right hand. When
the bird hit the ground, I gave the "fetch" command. Elijah bounded out,
hesitated over the "bird," and then picked it up and brought it back, head
high. If only I'd learned to shoot so quickly. Truth be told, had Elijah not
obliged on that October evening, my hunting career would have ended,
much as my earlier efforts had ended in frustration and disappointment.
This wasn't an opening day of the sort that gets much written about.

Teaching and family responsibilities gave me a couple of days to digest what had happened and what I had asked of Elijah. I was, frankly, not sure that I could become sufficiently good enough a wing shot to avoid the un-acceptable outcome of my first day afield. As inexperienced as I was at wing shooting, I was even more naïve about dogs. I had no way of knowing if Elijah and I had reached a lasting understanding on the basis of our first real day afield—our first day afield when the stakes were serious. Accounts of bird hunting I'd read glossed over experiences like mine. In the maga-zine articles, birds either fell from the sky stone dead or the dog was in-stantly to the rescue (of the hunter's conscience, I came to realize). There were few accounts of wing-tipped birds who hit the ground running and give even the best of dogs as tough a task as can be imagined. I'd expected everything to be unambiguous. As I reflect on it now, thirty plus years later, I was looking for a moral simplicity not all that removed from the packaged meats in the supermarket.

Three days later, my morning was open. The night before I set the alarm for 5:30. The forecast was for a clear but cold morning. Over the summer, I'd been rising early to work with Elijah before the heat got oppressive, so when the alarm went off, he was scarcely to be contained. He wasn't as troubled as I was. I fed him and our other dog, a splendid American fox-hound we'd named Houyhnhnm (in anticipation of breeding dogs, we'd decided on a Swiftian motif and named our imaginary kennel "Fourth Voy-age Kennels"), and took my time with my own toast and coffee, still not sure that I was up to what all too possibly would be a repeat of the ig-nominy of our opening day. Had Elijah not been as eager, and had we not logged all those hours together practicing for this very moment, I might well have rejoined my wife and slept in until it was time to get our oldest son breakfasted and off to school. Elijah paced between the kitchen table, where I sat contemplating, and the back door. Damn it all, I did not want to cripple another beautiful bird. But I had deliberately brought Elijah to the very edge of perfecting what decades of breeding had prepared him for. I was trapped. If I went back to bed, I would betray Elijah, whom I'd sys-tematically led to believe was my hunting partner. With some misgiving, I went to the basement to don my hunting garb and get my gun out of its locked case. When I returned to the kitchen, gun case in hand, Elijah could scarcely keep from crashing into the door. I realized I was hoist on my own petard.

Had I been slower opening the door to the van, Elijah might well have broken his neck in his eagerness to get into his travel crate. (Interestingly, once inside the crate, he was calm, as if sedated.) Off we went. I chose a covert about a half hour's drive away, as much to give me time to air my second thoughts as to maximize chances of success. We were now three days into the pheasant season, and as I came to learn over the years, all the easy birds were already in freezers or were half-devoured carcasses of the sort that Elijah and I worked with a few days earlier. The pheasants who had survived opening day had become quite evasive, and it took Elijah well over an hour to locate the first bird. He got "gamy"—the term of art describing the dramatic shift in a hunting dog's demeanor when scent of game is encountered—and after moving a few tentative steps, tail wagging furiously, he locked up. I was instantly bathed in a flood of conflicting emotions. I was, of course, thrilled (even nonhunters who have come along with me to see what hunting is all about cannot help but be excited by this moment). I was worried that either Elijah or I would botch things the way we had on opening day. I forced myself to go through with it, and when a hen pheasant flushed I was relieved (in those days, only cock pheasants were legal). I relaxed as I watched her glide into a hedgerow at the far end of the field we were working through. For a moment, I overlooked the fact that Elijah was still intensely on point. I turned around to face him and from behind me heard the thrashing of wings and the cackle of a cock pheasant. Whirling around, I found myself almost face to face with the bird as it rose out of the knee-high grass. To this day, I can still see that bird etched against the sky, long tail trimmed for flight and its ringed neck stretched out. It took me a moment to gain composure and then, all ambivalence evaporated, I brought my gun to my shoulder, swung through the flight path of the bird and squeezed the trigger. The bird crumpled and fell to the ground in thick swale—where it would be virtually impossible for me to find. I gave Elijah the command "fetch," and held my breath, which was not an easy thing to do because my heart was pounding with excitement.

Elijah made a beeline for the spot where the bird had landed, bounding high with every stride to rise momentarily above the tall grass to make sure of his destination. By the time I caught up with him, he was creeping forward, tail again wagging furiously, with his nose fairly scraping the matted grasses. He lunged forward, his head disappeared beneath the mat, there was a brief scuffle, and he backed out with the rooster held tightly but gently between his jaws. Head high, as if to say "no more of those lousy carcasses you made me fetch the other day," he brought the bird to me and deposited it in my outstretched hand. I was laughing through the tears that streaked down my cheeks as I laid the bird on the ground and reached to hug Elijah. I had just enveloped him in an embrace when the bird bolted and buried itself under the grass. Elijah wiggled free and quickly recaptured the bird, our first pheasant. In the ten years we hunted together, Elijah failed only once to retrieve a bird I'd shot. The ethical circle had been closed: Elijah, the birds we hunted, and I formed a universe in which we each had a part to play. It seemed then, with all the pieces in place for the first time, a sensibly moral universe. It still seems that way to me.

The drama Elijah and I and the game birds we pursued enacted has a timeless quality to it. Hunters have been chasing birds with dogs for hundreds of years. This history makes it tempting to argue, as indeed many have, that hunting is instinctual, rooted deeply in our evolutionary history. I have no doubt that this is so. I also am convinced that our evolutionary history is largely irrelevant to an understanding of hunting. Yes, our remote ancestors were hunters and our evolutionary development gave us the tools, especially the tools to make tools, that enabled us to be predators and rewarded those who were good at it. Elijah's genes were consciously selected, beginning well before anyone knew about genes, to enable him to be an eager partner in the hunt. But genes and instincts alone, neither mine nor Elijah's, made us hunters. Elijah would have been happy as a pet. To be sure, he would have pointed robins and butterflies and would have brought back various animals, vegetables, and minerals to us: pointing and retrieving were in his genes. It's a good deal less certain what genes I carry that inclined me to hunting, but I dare say the answer is either "none" or "all of them." Neither answer is very helpful. Had I decided to quit hunting after that demoralizing opening day, I would not have been going against my nature or defying my genes. I would have simply decided that hunting was not "my thing" and gone about living my life somewhat differently. From the moment the first human killed an animal for its flesh, hunting became as much a culturally defined as it was a biologically ordained activity. Over the millennia, layers and layers of culture have been deposited on whatever genes have contributed.

Elijah and I were not reenacting some primordial scene those October and November mornings. We were following a script whose first draft is ancient and whose subsequent revisions form and inform the experience of hunting. Indeed, the particular script we were following is a revision drafted in the late nineteenth and early twentieth centuries, primarily in the United States. This version of the hunting script introduced the notions of "fair chase" and restraint as well as an explicit emphasis on a humane (i.e., quick) kill. It was this later standard I had failed to meet and which had demoralized me on my first hunt with Elijah. I wasn't disappointed at the prospect of coming home empty-handed. My masculine pride, whatever that might be, wasn't injured. I had violated an ethical obligation I had voluntarily undertaken when I bought a shotgun and a gun dog puppy. No one was there to witness the debacle and even had there been witnesses, I would not have been reproached. They would have known that I felt bad.

On that morning in late October, whatever doubts, including self-doubts, I had harbored about hunting evaporated. In the three decades since Elijah and I began hunting pheasants, the number of hunters has declined, especially in New England, and objections to hunting have steadily gained currency. By the late 1990s, it was clear that hunting was headed for trouble. It was time for me to revisit all those personal ambivalences that Elijah and I had dispelled years earlier. The book that follows is not simply another defense of hunting, though there's no blinking the fact that hunting needs defenders. My goal is less to defend than to try to understand the place hunting occupies in the contemporary world. Why are people drawn to hunting and what, if anything, changes when a person picks up a gun or a bow with the intention of killing an animal? By what right do hunters kill animals? Mortal stakes.

I

FROM VILLAINS TO HEROES—
TO VILLAINS?

For much of our nation's history, attitudes toward wildlife alternated between indifference, fear, and utilitarianism—if an animal was valuable for its meat, skin, or feathers, we hunted or trapped it. If it threatened us, our livestock, or our crops, we killed it. If it was neither valuable nor threatening, we said "to hell with it." Appreciation for animals for their own sake, because of their beauty or gracefulness, began in earnest in the middle of the nineteenth century, fueled, ironically, in part by a uniquely American development—sport hunting. Audubon, a hunter, was also a naturalist and, of course, an artist. His lifelike renderings of birds native to North America helped to define the beginnings of this shift in thinking about wildlife. In Audubon's day, wildlife was under heavy siege. Market hunters were shooting, netting, and trapping virtually every critter that could be sold. Predators were trapped, shot, and poisoned in the mistaken view that their elimination would mean the domestic animals and wildlife we valued would thrive.

After the Civil War, the pressures on natural resources, including wildlife, intensified as industrialism took off and our population swelled, boosted by a steady stream of immigrants from Europe and Asia. Wildlife that was not directly exploited got hammered by the loss of habitat that resulted from our logging, mining, plowing, and damming. Had these practices continued, Audubon's paintings might well have been all that was left of our once abundant diversity of wildlife. Audubon's paintings helped Americans to see something other than nuisance or commercial value in wildlife, but as importantly, he was an active promoter of hunting not merely for profit or food but, rather, for the intimate contact with nature that hunting afforded.* At the risk of oversimplification, it is as though attraction to the wild and appreciation for unspoiled nature grew in proportion to our taming of wild nature. As cities grew from towns and fields and meadows replaced forests, Americans began to seek solace and even salvation in immersion in nature (Albanese 1990; Schmitt 1969).

This growing regard for nature began to coalesce into what came to be called the "conservation movement." Spurred by a growing realization that the unrestrained exploitation of natural resources was jeopardizing wildlife and threatening to upset what some were coming to regard as the balance of nature (Worster 1977), the movement began gathering momentum shortly after the end of the Civil War. Among the first fruits of this shift in popular sentiment were the creation of national and state parks, Yellowstone, the first and most famous (1872) of the former; and New York's Adirondack State Park, created by a series of laws between 1885 and 1894, whose size (six million acres) exceeds the combined expanse of Yellowstone, Grand Canyon, and Yosemite Parks (Schneider 1997). When Theodore Roosevelt carried the cause of conservation with him to the White House in 1901, "conservation" became a household word. Large blocks of land were acquired by the federal government, primarily in the still sparsely settled West, some designated parks to be managed by the newly created National Park Service and others assigned to the stewardship of the also new U.S. Forest Service and the U.S. Bureau of Land Management. States developed their own parallel agencies and began establishing state parks and forests.

Central to this shift were a group of men, most of them wealthy, who were avid hunters and anglers, men who shared Audubon's love of nature as well as his passion for "blood sport." Appalled by the drastic decline in game caused by a combination of habitat loss and market hunting, these men began concerted efforts to protect wildlife from commercial exploitation. They founded magazines to promote the virtues of sport hunting and the ethic of fair chase, as distinct from market or "pot" (subsistence) hunt-

---

*This brief account of the origins of sport hunting and its intimate connection with a broader shift in attitudes toward nature that gave us the idea of conservation is based on the work of Reiger (1975), Trefethen (1975), Herman (2001), and Nash (1982).

ing. They also formed organizations, notable among these the Audubon Society, to promote the broader goal of conservation. The Boone and Crockett Club, founded in 1887, was created to press for legislation that would end market hunting and the unregulated taking of fish and game. Club member and congressman John Lacey promoted the first federal game protection law, the Yellowstone Park Protection Act of 1894 (Organ et al. 1998), and later an act named for him, the Lacey Act (1902), which essentially ended the commercial traffic in wildlife. States followed suit by enacting laws regulating the taking of fish and game. Hunting and fishing seasons were established, with an eye toward protecting fish and game species when they were reproductively active or otherwise unsportingly vulnerable, and bag limits were imposed to further relieve pressure on wildlife populations.

The overarching goal uniting these efforts was the promotion of an ideal of sustainable use of natural resources as well as the protection of some areas from commercial exploitation. Employing scientifically grounded forestry and agricultural practices as well as the newly emerging sciences of ecology and wildlife management, men like Gifford Pinchot, Roosevelt's pick to be the first head of the U.S. Forest Service, were sure that it would be possible to arrive at determinations of harvest levels that would ensure a steady supply of natural resources far into the future. Selective cutting of our forests, crop rotation and improved tilling methods, game laws that would ensure populations robust enough to replenish themselves after the fall's hunt, and land set-asides that would remain free of the commercial pressures to maximize yield, would, they thought, conserve our natural resource base. The conservation movement linked conservation to ensuring national self-sufficiency, which would, in turn, help us avoid alliances and trade relations that smacked of colonialism and foreign domination. Appealing to both nationalism and nature appreciation, the movement effected a change of heart.

To be sure, the conservation movement had to contend with much resistance to this new ideal. The first challenge was relatively easily turned aside and would scarcely be worth mention were it not for the fact that it has resurfaced and now represents a more serious challenge to the ideal of conservation. Conservation is premised on seeing nature as a bundle of resources that we should use prudently. Even things that have no apparent use are nevertheless linked to other things that are useful, and so we should regard nature, as a whole, as stewards. For a small but articulate and inspirational group, most notably John Muir, the idea that nature was a mere pile of resources was unacceptable. Muir saw divinity in nature, and he wanted as much of it set aside as possible. The growing fascination with nature meant a growing market for books and essays on nature themes. Though there were plenty of solid contributions to what we now call "nature writing," among them Muir's, there were also lots of writers and readers whose enthusiasm for nature overwhelmed their commitment to the truth. Dubbed "nature fakers" by the leading nature writer of his day, John Burroughs, writers like Ernest Thompson Seton, William J. Long, and Jack London were busily promoting a view of nature that stressed preservation rather than conservation, and attributed to animals all sorts of capacities that made them appear almost human (Lutz 1990). Their tales made for good reading but the literary liberties they took finally strained all but the most credulous. Their allegations simply could not withstand the blistering barrage of criticism from Burroughs and other naturalists, including Theodore Roosevelt. Excesses of nature loving were pushed further and further to the side as public policy moved steadily to embrace conservation.

Other challenges proved harder to overcome and have continually dogged the heels of conservationists. The early conservationists had to contend with ignorance, their own and the public's. With the best of intentions, but all too often with faulty or incomplete understanding of how nature works, forests were destroyed, watersheds disrupted, wetlands erased, and the range overgrazed, all with the imprimatur of "conservation."* To compound the blunders of the experts, those owning or working in extractive industries like logging were unhappy, to put it mildly, with the constraints a conservation regimen imposed. Farmers were slow to adopt new practices they feared would raise the costs of production. From the perspective of people who lived on the land, conservation sounded like an assault on their way of life, just as, today, attempts to limit grazing on public lands or road building in our national forests are resisted by the ranchers and loggers who fear for both their livelihoods and their way of life (Russell 1993; Brown 1995). On the ground, far from the social clubs, legislative hearing rooms, and government offices, practices were slow to change.

Conservation was an ideal all too often honored in the breach. We continued to clear land at a ferocious pace and log our forests at rates no one could claim were sustainable. The same can be said for our mining, agriculture, and commercial-fishing practices. Wild animals remained a necessary source of

---

*For a detailed account of how this played out in one area in Oregon, see Nancy Langston, *Forest Dreams, Forest Nightmares* (1995).

food for many in the nation's small towns and rural areas, and proceeds from traplines were by no means trivial to many household economies. Wildlife also continued to directly or indirectly compete with our domesticated animals for food. Harvesting deer or beaver or birds was as much a part of agrarian life as planting crops and cutting firewood. So was the killing of "varmits": many animals made life harder than humans were willing to accept, and so they were shot or trapped even if they were not consumed. Well into the twentieth century, a rifle or shotgun lay at the ready in most rural households against the possibility that a fox or a hawk would come after the chickens.

By contrast to others who were directly dependent upon nature's bounty, hunters and anglers rather quickly embraced the conservation ideal. To be sure, as Warren (1997) makes clear, rank-and-file hunters often resented the imposition of license fees, bag limits, and other constraints that limited the taking of fish and game, and many hunted outside the law. Even now, some still do. But the conservation ideal, including the ethic of fair chase, rapidly gained the high ground. This change of heart and behavior was reinforced as the benefits of this new regime began to become apparent. Game and fish stocks began to recover. Revenues generated from license sales supported state agencies that launched scientific studies aimed at improving our ability to manage game species to ensure their abundance. The newly created U.S. Fish and Wildlife Service played a vital role in advancing scientific knowledge of wildlife and habitat restoration and maintenance. Federal and state money was also devoted to land acquisition to ensure public access to quality hunting and fishing opportunities. In the 1930s, these efforts got a major boost by the passage of the Pittman-Robertson Act, which levied a tax on all hunting-and-fishing-related products, the proceeds of which were dedicated to enhance the conservation of fish and game through land purchase, the study of wildlife and wildlife habitat, and the continuing promotion of the conservation ideal through state-run programs of hunter education. By the mid-twentieth century, the ethos of sport hunting and fishing had thoroughly permeated the ranks of the nation's hunters and anglers and both activities were becoming increasingly popular recreations. Indeed, the number of hunters swelled to embrace nearly one quarter of the nation's men by the 1950s.*

This is really quite a remarkable story, all in all, as much for how hunting

---

*Until recently, few women hunted. I explore the social characteristics of hunters, including gender, in chapter 2.

21

was transformed as for the spectacular recovery of many game as well as nongame animals that can be directly linked to the concerted efforts of hunters and anglers. Indeed, I think it is fair to say that had the conservation ideal been as thoroughly embraced by farmers, commercial fishers, and our extractive industries as it was by hunters, we would not be facing anything like the environmental disasters that now loom. The promotion of conservation went hand in hand with the promotion of the ideal of the sport hunter—a person who understood the need to practice self-restraint, a person who took pains to become knowledgeable about wildlife and nature more generally, a person who was respectful of nature as well as of the safety and well-being of others. In short, the hunter was transformed from a rural half-savage into an upright, middle-class figure. As we grew more and more urban and industrial, hunting, along with the militia, came to be regarded as bedrock American virtues, the very things that made us free and democratic. In his fascinating and detailed study of the history of hunting in America from colonial times to the early twentieth century, Herman (2001) argues that the elite hunters who played a decisive role in the conservation movement in effect helped create an American ethnic identity, a core feature of which was the celebration of a highly romanticized version of the American hunting tradition.*

Even though most Americans are but a generation or two removed from what might be called the heyday of hunting, a time when hunting was broadly engaged in and even more broadly accepted, hunting has clearly fallen from the pedestal. As I show in detail in the next chapter, rates of participation in hunting have declined dramatically from the levels attained in the 1950s and 1960s. As the number of hunters declines, criticism of hunting has intensified. What has happened?

"It's payback time for animals."**

Across a broad range, wildlife is staging a comeback in North America. Though there are many species, especially the small creepy-crawly ones, that

---

*I return to this idea in chapter 6.

**This is a quote from an unnamed participant in a conference I attended in the fall of 1998 that was devoted to considering the prospects for reintroducing wolves to the Adirondacks. The speaker was insisting that we should at long last be willing to put up with whatever damage animals might cause us, given how much abuse animals have endured at our hands.

have not participated in this happy tale of recovery, it is nonetheless one of the lamentably few bright spots in recent environmental history, a history that more often records loss than recovery. Some of this recovery has been the result of deliberate interventions like the reintroduction of wolves to Yellowstone and the passage of the Endangered Species Act (1972). But we cannot take credit for all the good news. The resurgence of some species has been more spontaneous, the result of coincidence as much as deliberate promotion. The white-tailed deer is the archetype of this form of resurgence. Hunted to extremes in early colonial times, pushed back even further by logging and the clearing of much of its range for agriculture, white-tailed deer reached historic lows early in the twentieth century. Then, state game biologists intervened to regulate hunting pressure and to protect does. As importantly, after World War II, Americans began moving in droves to the suburbs. Farms were turned into subdivisions, and in between the asphalt and manicured lawns, green belts grew in crazy-quilt patterns. This made for ideal white-tail habitat. In more remote wooded areas, the building and population booms of the fifties and sixties unleashed another wave of intense logging, in the wake of which extensive new growth also well suited the white-tail deer.

Unaware of or indifferent to the effects our changing lifestyles were having on wildlife, we have been creating what ecologists call "edge" over much of our landscape: the dense tangles of brush, swale, and young saplings that appear as disturbed land, whether from natural or human-inspired causes, begins to recover. The continuing collapse of the small family farm has added even more edge as once-grazed pastures and tilled fields revert to brush and then to early successional forest. For those species that depend on the young, tender shoots of new trees and the insects and small mammals that are drawn to the tangles of grasses, shrubs, and vines that characterize edges, these shifts in our economic and residential patterns have been a godsend. To be sure, not all creatures thrive in this sort of disrupted and fragmented landscape— creatures like the spotted owl, which require large and unbroken deep forest, are pushed to a different kind of edge, the edge of viability, by these changes. Many migrating songbirds are experiencing declines at least in part because the unbroken habitat they require is now pockmarked by subdivisions and criss-crossed by highways, high-tension lines, and interstates.*

---

*This sort of trade-off is axiomatic. There is almost nothing we can do to create a general abundance or diversity of wild things. Where such an abundance exists, we may be well advised to leave well enough alone. But over much of our landscape, most of which has been

Still, for most of the species we first think of when we think of wildlife, the last few decades have been quite good. Indeed, it is not stretching things to say that we are facing an embarrassment of riches. Even though our spirits rise with news of the steady growth and dispersion of wildlife, problems have begun to arise. More importantly, proposed solutions to these problems have become the focus of steadily intensifying and increasingly vituperative debates (Dizard 1999; Nelson 1999). White-tailed deer are scourge to suburban gardeners and a hazard to motorists. Beaver are raising water tables and in turn flooding basements and impairing septic systems as well as leveling the wooded buffers that provide suburbanites with a sense of privacy and, ironically, closeness to nature. Coyotes are augmenting their customary diet of rabbits and mice with the pets that roam our small towns and suburbs, creating grief and anguish in their wake. Canada geese have found expansive lawns, playgrounds, and golf courses accommodating, so much so that large numbers of them have decided to stay put and forgo the rigors of annual migration. All across southern New England and the mid-Atlantic states as well as in parts of the Middle and Far West, these resident geese are creating both nuisance and public health concerns. Lyme disease, carried by a tick that uses the white-tailed deer as host, has become endemic across large sections of the country and poses serious health risks to those exposed.

Hunting and trapping have been the traditional responses when wildlife causes damage, inconvenience, or risk to humans. In recent years, however, the ranks of hunters and trappers have been declining, and public disapproval of hunting, and even more of trapping, has increased. Indeed, hunting and trapping have become targets of a growing and ever more sophisticated animal rights movement committed to, among other things, putting an end to these activities.* The result has been a growing list of communities forced to agonize over how they might—or whether they even have a right to try to—manage wildlife populations in their midst. One recent report of a dispute in a community on Long Island will suffice to give a flavor of controversies being played out across the country.

---

cleared at least once as well as felt the slice of the plow or the hooves of livestock, whatever original diversity there was has given way to the predominance of those flora and fauna that are able to accommodate to the ways we have modified the environment.

*As of this writing, one of the most prominent of the animal rights organizations, People for the Ethical Treatment of Animals (PETA), has announced that they are adding sport fishing to the list of exploitative uses of animals to which they wish to put a stop.

Like many of its neighboring townships, North Haven is home to a deer population that is almost as large as the human population.* Collisions with deer are an ever present danger. Gardening is out of the question, unless the gardeners are prepared to erect a substantial electrified fence around their gardens. Ornamental trees and shrubs are a worthless investment. A survey of town residents found that "more than half had had Lyme disease." One woman reported that her husband had had the disease four times and she has contracted it twice. The woods that gives the town its natural feel are feeling the pressure too. "'The deer are also destroying the understory, or low-lying vegetation of the woods,' said Bob Ratcliffe, the village mayor. 'We don't have seedlings to replace the dying trees.'"

The village appointed a deer management committee which recommended a controlled hunt, a plan the village board adopted. In the year since the plan was approved, a hundred deer have been shot. But controversy continues to swirl around the deer. The issue has divided neighbors. One resident spends "hundreds of dollars a week on food for deer." The family next door has erected a barricade to keep the deer out. Many villagers reject the idea of killing deer, and they have filed suit to stop the hunt. Why are so many people suddenly willing to put up with wildlife?

Wild animals now occupy a very different place in our lives. The Cartesian divide between humans and the rest of creation is no longer clear or intuitively obvious. A 1993 *Los Angeles Times* poll found that 47 percent of the adults polled believed that animals are "just like humans in all important ways." The University of Chicago's National Opinion Research Center found, in their 1994 General Social Survey, that 28 percent of those polled thought that "animals should have the same moral rights that human beings do."** Even those who remain unwilling to collapse the

---

*This account is based upon a *New York Times* news story, March 24, 1996, p. 41, and upon a more in-depth account of the North Haven situation set in the larger context of the growth in suburban wildlife populations sketched above: Anthony Brandt, "Not in My Backyard," *Audubon*, September–October, 1997, p. 58–62, 86–87, 102–03.

**I rely heavily on the survey research conducted by the National Opinion Research Center (NORC), particularly in the next chapter. Beginning in the early 1970s, they conducted an annual survey, called the General Social Survey, in which a carefully selected random sample of 1500–2900 adult Americans were asked an array of questions about their attitudes and behavior regarding a broad range of issues and concerns, from stances on highly charged issues such as abortion to intimate details of their personal lives. Since 1994, NORC has conducted the General Social Survey every other year. I rely most heavily on data from the 2000 survey, supplemented, where necessary, by data from earlier surveys done in 1994, 1996, and 1998.

distinctions between humans and animals are likely to see wild animals less as a threat or force with which to contend than as a symbol of wildness they wish to preserve. Though there are few among us who would go to the extreme of spending hundreds of dollars a week to feed deer, there are obviously many who are now willing to put up with inconvenience and even some risk of disease or injury in order to see wild animals or at least be comforted by knowing that wild creatures are back. In the face of so much bad news about the environment, the honking of a wedge of geese flying overhead or the flash of a red fox coursing along the edge of road can be deeply reassuring—even if the geese are headed to the golf course instead of remote northern breeding grounds and even if the fox might be rabid.

Since we are no longer locked into a close and direct dependence on the land, wild creatures no longer present themselves as either threat or competitor. To be sure, police switchboards light up when a bear or a mountain lion is spotted in a residential area, and animal control specialists do a brisk trade relieving homeowners of skunks nested under the deck or squirrels in the attic. But such encounters and annoyances do little to damage the reputation of wild animals. This regard for wildlife is not simply a reflection of urban-suburban thinking. Rural folk, too, have begun to turn away from the traditional disparagement of wildlife. Increasing numbers of ranchers are willing to explore ways of protecting their flocks and herds without resort to traps and poisons. A recent survey of Vermont farmers revealed that even among those farmers who had experienced losses due to wildlife, nearly half were opposed to stepped-up efforts to manage the species involved (Duda et al. 1997). Of course, attitudes differ as one moves from the farm to the city or suburb, but the differences appear to be much smaller than they once were.

Indeed, even when risk is great, public tolerance for wildlife remains high. The steady spread of Lyme disease has done little to alter affection for white-tailed deer, despite the growing evidence that the disease is more serious and variants of it harder to treat than has been thought. It is worth noting that the woman in North Haven who, with her husband, had contracted Lyme disease several times was nevertheless opposed to the controlled hunt. Similarly, the outbreak of rabies that moved northward over the past two decades and is now endemic among raccoon populations throughout the Northeast and mid-Atlantic states has scarcely caused a ripple of concern beyond those entrusted with protecting public health. In fact, in Massachusetts, where rabid raccoons have been confirmed in virtually every community in the state, including Boston, animal rights activists successfully campaigned for a referendum outlawing, among other things, the use of leg hold and body-gripping traps, still the most effective instruments for the control of fur-bearing mammals. The referendum passed by a two-to-one margin. A generation ago, there would have been a hue and cry urging the aggressive trapping of raccoons to prevent the spread of rabies. In the same vein, in the spring of 1996 voters in California passed by a comfortable margin a proposition to end all hunting of mountain lions, even though two women, one in Southern California and the other in Northern California, had recently been killed by mountain lions.

This enthusiasm for wildlife is part of a larger shift in Americans' attitudes toward the environment, a shift that seems more and more to involve idealizing nature and disparaging human interference with nature. Again, data from the National Opinion Research Center is instructive. The 1994 survey mentioned above also revealed that half of all respondents agreed that "nature would be at peace and harmony if only human beings would leave it alone." Forty-eight percent agreed that "almost everything we do in modern life harms the environment." Management of natural resources is also suspect. A group of anthropologists recently found that most people they talked with held environmental management in low regard. They regarded interventions in the environment more part of the problem than a possible solution to our environmental woes (Kempton et al. 1995).

The opponents of hunting deer in North Haven, like the protectors of mountain lions in California or the people who oppose trapping beaver, are, it would seem, far from oddballs. While they may not be a majority, their views cannot be thought kooky or on the fringe. Layered through the disputes over the management of "problem wildlife" are images of hunting and hunters that are fed by our deepest feelings about our relationship to nature and wildness as well as our attitudes toward killing. Even as we seek to escape the asphalt and concrete with which we have demonstrated our dominance over nature, even as our hearts are lifted by the accounts of wildlife long absent from our landscape staging remarkable comebacks, the ways we have distributed ourselves over the landscape and otherwise modified the natural environment means that we will have conflicts with wildlife. It also means that we will disagree with one another about how to live with wildlife. This means hunting will be in the spotlight.

27

The Problem with Hunting

In view of such increased regard for wildlife, it is not entirely surprising that the stock of hunters has fallen. In one of the earliest and careful studies of attitudes toward hunting, sociologist Stephen Kellert (1978) found that one-third of the public favored an outright ban on hunting. Opposition to hunting nearly doubled when people were asked about sport or recreational hunting. Only hunting in order to put meat on the table gained broad support. In a more recent poll (1993) by the *Los Angeles Times* referred to earlier, a majority of respondents opposed sport hunting. Hunting eked out a slender supportive majority among men but two-thirds of women polled disapproved. Sixty percent of young people (under 24) disapproved, suggesting that support for hunting will decline even further over time. Of course, the relationship between regard for animals and attitudes toward hunters is complicated. The reputation of hunters has long been problematic, not least because a wide range of activities, some of them quite repellent, get folded into the rubric "hunting." So, for example, the anthropologist Matt Cartmill (1993) describes passing two scruffy men wrestling with what appeared to be a mattress on the side of the road leading to his farm in North Carolina, a place where he and his wife had been enjoying the comings and goings of a doe and her two fawns. His account continues: "The two men had left by the time I started back home, so I stopped to look at the mattress, which turned out to be something wrapped in a bloodstained pink sheet. I threw back the sheet and discovered the decapitated, skinned, and gutted body of a small deer. . . . I have thought a lot about those two *hunters* and their motives since then" (227–28; emphasis added).

How would the force of this passage been changed had Cartmill called the pair "poachers" instead of "hunters"? Is there any reason to distinguish between hunting and poaching? Certainly from the animal's point of view, there is precious little difference. But clearly from the point of view of humans, there is considerable difference between hunting and poaching. With the steady rise in solicitude for wild animals, however, it may well be that the distinction between poaching and hunting is collapsing in the minds of the general, nonhunting public.

Most accounts of hunting assume that it has long held an honored place—hunting was indispensable to human well-being, and those who were skilled at it were highly respected members of their communities.

Hunters were strong and intrepid. They were also acutely perceptive and, at their best, as gifted at sorting and sifting for clues as any great detective. There is little doubt, also, that until very recently, prowess in hunting translated rather smoothly to the skills and aptitudes required for both defensive and offensive war making, adding elements of bravery and heroism into the mix of lore about hunting.* All of this adds up to the hunter being a central figure in the community. Indeed, as Joseph Campbell's (1988) sweeping survey of mythology makes clear, myths and legends of the hunt are staples of all cultures and raise the hunter above mere mortals.

The literature of hunting routinely invokes this long thread of veneration and draws from it all sorts of claims of how hunting forms character, heightens the keenness of the senses, and teaches humility. Indeed, the brilliant and provocative Paul Shepard (1973) has pressed this to the conclusion that most modern woes trace back to the moment we left hunting behind in favor of farming and pastoralism. The most famous American hunter, Theodore Roosevelt, promoted hunting as an activity that would keep the nation's men intellectually and physically vigorous. Fearing that the extreme specialization and division of labor entailed by industrial growth would render men weak and irresolute, he urged upon his countrymen a "vigorous life" of wilderness camping, hunting, and outdoor adventure.** Unlike most politicians, he practiced what he preached, and his example remains a remarkable one, no matter what one might think of the values he promoted.

Whatever hunting might or might not do for character building, recent work, most notably that of historian Daniel Justin Herman (2001), suggests that hunting was not nearly as integral to life in colonial America as has commonly been thought. The English were committed to establishing an agrarian society and to rendering what Bradford referred to as a "howling wilderness" into a pastoral civilization befitting God's plan. The work entailed in this would leave precious little time for hunting. Indeed, it seems probable that much of the game that graced the colonists' tables was secured by trade with the natives, whose bows and arrows and hunting

---

*For an insightful and provocative discussion of this, see Ehrenreich 1997.

**In fact, though Roosevelt was concerned with manhood, numbers of eastern middle-class women, many of whom having had bouts of depression, headed to the Southwest and the Rocky Mountain states to live their own version of the robust life. For one such redoubtable woman's story, see Peggy Pond Church, *The House at Otowi Bridge* (1960).

skills were far superior to the English guns and know-how (Martin 1978; Krech 1999). The exceptions to this seem to have been fowl and large predators. Birds were considerably easier prey than were larger mammalian species, if for no other reason than there were large and concentrated numbers of pigeons, doves, turkey, and waterfowl. Moreover, some bird species were nearly tame and could be taken more easily by a thwack with a stick than with a gun. Deer were not nearly so cooperative. Thus, the fabled first Thanksgiving was made up of fowl gotten somehow by four Puritan "hunters" and five deer contributed by the Wampanoags (Herman 2001).

The other exception, predators, were relentlessly pursued. Wolves and mountain lions simply had no place in the colonists' pastoral imagination. In 1630, Massachusetts Bay Colony instituted a bounty on wolves, which it had to suspend briefly because the incentive proved so effective that the policy quickly became a financial drain on the Colony's coffers (Wagner 1982). Birds, too, were a scourge on the grain crops the settlers were trying to establish, and so the shooting of birds was also encouraged. We may never know whether it was the natives or the colonists who did the killing, but the flesh, hide, and fur of wild animals constituted a significant chunk of the net value of exports from the colonies for a considerable portion of the seventeenth century (Cronon 1983). In this sense, hunting was understood to be important, and the skills of the hunter, whether native or English, were no doubt appreciated.

Beneath this heroic narrative, though, there is another story line. The good people of the Bay Colony, however much they appreciated venison and took comfort in securing their crops and animals from the creatures of the wild, also had to dodge errant musket balls and birdshot. Wagner (1982, 31) cites a 1678 report from Boston: "Several Persons have been killed by such as have pretended to have shot at Fowls, birds &c., and that in or near Highwayes, and many take the boldest upon them, Youths and grown Persons, too frequently to shoot within the Limits of Towns, Orchards, Gardens, &c." Hunting was associated with unruliness and what we would today call "hard living." In England, only the elite could hunt. Ordinary folk could and did poach, but the penalties were severe and the gains, if any, were few. Those who took the risk were hardly the most reputable. Moreover, the Puritans who dominated New England for decades, opposed as they were to the landed gentry and the Crown, were outspoken in their denunciations of virtually all features of English culture. As Herman (2001) notes: Puritans led efforts in England to end hunting, cockfighting, and the other blood sports by which the high and low entertained themselves, though it should be noted that the gentle souls who boarded the *Mayflower* had been admonished by their leaders to bring guns and be knowledgeable in their use, less to get game than to repel hostile natives (Zuckerman 1977). For the Puritans, hunting took men away from the activities that advanced the covenant with God: prayer, work, and devotion to community.

Puritans could not possibly think hunting was heroic. More likely, they saw it as a form of debauchery. Worse, as the report from Boston suggests, hunting was a threat to good order. This was so in several respects, probably the least important of which was the risk of injury. Hunting, insofar as one got good at it, could weaken the hold of the community on the individual. Armed men capable of putting meat on the table by themselves might be less inclined to follow the strict rules and submit to the close scrutiny the Puritans insisted upon. With abundant land and a seemingly inexhaustible supply of wildlife, the temptation to strike out on one's own must have been an ever-present worry for the religious leaders of the Bay Colony. Fear of the unknown and of the natives no doubt held people in check, but these fears surely dissipated over time. In this context, it is easier to see why the colonists might, all in all, prefer to leave hunting to the natives, who were "savage" to begin with, rather than encourage Englishmen to sink to the level of the natives.

The English who settled in Virginia and further south did not come here to repudiate English manners and took more eagerly to hunting. There, the debate was whether hunting should be a popular as opposed to an elite activity. The matter was settled in favor of open access, which introduced its own set of aggravations that kept hunting from being warmly embraced. As land got cleared and crops planted, conflicts between hunters and landowners appeared and have remained with us down to the present. As the colonists, south and north, busily carved up the landscape into private holdings, trespass became a problem along with determining who had rightful claim to wildlife: were deer on my property mine or could anyone come shoot them? As James Tober's (1981) work makes clear, such questions as these were the subject of much legal wrangling and legislation. Laws are one thing, enforcing them is quite another matter. In the grip of the chase, hunters were not likely to pay close attention to the niceties of property lines, and in their enthusiasms they could disturb livestock (or worse) and damage crops.

Herman (2001) convincingly shows how hunting was thus held in low regard for much of our early history, even though hunting (and trapping) grew in importance as whites pressed ever westward. Hunting began to rise in stature, Herman suggests, because it was intimately linked to the fulfillment of the national ambitions that got codified in the doctrine of Manifest Destiny. The exploits of Lewis and Clark and larger-than-life figures like Daniel Boone fired popular imagination and laid the basis for what, by the last third of the nineteenth century, became a national narrative of progress and growth in which hunting played a central role and hunters were the intrepid advanced guard of civilization. Gone were the sorts of objections to hunting that Benjamin Rush enumerated in 1790 (Herman 2001, 65–66):

1. It hardens the heart, by inflicting unnecessary pain and death upon the animals.
2. It is unnecessary in a civilized society, where animal food may be obtained from domestic animals with greater facility.
3. It consumes a great deal of time, and thus creates habits of idleness.
4. It frequently leads men into low, bad company.
5. By imposing long abstinence from food, it leads to intemperance in eating, which naturally leads to intemperance in drinking.
6. It exposes to fevers, and accidents. The news-papers are occasionally filled with melancholy accounts of the latter, and every physician must have met with frequent and dangerous instances of the former, in the course of his practice.

Americans were dealing with very ancient themes as they wrestled with hunting. At the core of this problem is one disarmingly simple fact: hunting is *exciting*. In its earliest, most primitive forms, the excitement no doubt arose from the ambiguity of just who was hunting whom. When technology advanced sufficiently to give humans an edge on the animals, the excitement began to derive less from risk than from the challenge of stalking or chasing and the powerful emotions that are unleashed by the kill. Excitement, whatever its origin, makes impulse control difficult and the exercise of good judgment more tenuous. And then there is the killing.

There are few axioms in the study of human culture and behavior, but one of them surely is that where we find myth and ritual, anxiety and ambivalence are near at hand. Scholars of myth, like Campbell, and anthropologists (Nelson 1990; Brody 1982, 2000) give us a rich record of the lengths to which our forebears went to circumscribe killing, whether of animals or of humans. "Primitives," no less than the most ardent of animal rights advocates, understood that taking a life is serious business, even if the life taken is that of a "mere animal." Even when solicitude is not directed at the animal, killing raises doubts about the killer's psychological stability. If you can bring yourself to snuff out the life of a creature, what is to keep you from doing the same to me? The question grows more disturbing when the hunter acknowledges a certain pleasure in the hunt. The mythic figure, Orion, the Hunter, underscores precisely this anxiety (Bergman 1996). Orion meets his end, stung by a scorpion, as a result of his attempt to rape Artemis, the goddess of the hunt. Character building through hunting is a fragile enterprise, it would seem.

Among all the other things they do, myths and legends of the hunt serve to frame or bound the activity, much as rules of engagement serve to define what is legitimate use of lethal force in warfare. Our ancestors may have extolled the virtues of hunters as much to exhort them to restrict the killing to culturally acceptable creatures as to honor them. At the very least, honor comes mixed with fear. Not surprisingly, contemporary critics of hunting echo Rush's charge that hunting hardens the heart of the hunter and makes him more violence-prone. Merritt Clifton, editor of the publication *Animal People,* has pressed this view vigorously. In one article Clifton prefaces consideration of data purporting to reveal a relationship between hunting and child molestation with the following descriptions of two serial killers: "[A] team of 165 volunteers shoved snow from the frozen forest floor near Raquette Lake, where *hunter* Lewis Lent, Jr. said he'd killed and buried 12-year-old Sara Ann Wood. . . . Arthur Shawcross, the most notorious serial killer of recent years in the Albany region, was also an *inveterate hunter*" (Clifton 1994, 1; emphasis added).

Not bothering to inform readers that Lent's mental functioning was well below normal or to define what it means to be a hunter, much less an "inveterate hunter," Clifton proceeds to report that in a county by county analysis of New York, there is a correlation between the proportion of residents holding hunting licenses and the frequency of reported incidents of sexual assault on children. Clifton's data are suggestive; the question is, "suggestive of what"? To show a relationship between hunting and sexual abuse, one would need to compare *hunters* and *nonhunters,* not counties.* Clifton knows this but ducks: "if the population of hunters and pedophiles not only parallel but overlap, hunting might no longer be just a common element in the backgrounds of most sexual predators: it might begin to be recognized as a symptom of sexual abnormality in and of itself" (7). Clifton makes some huge leaps in this short passage and invites readers to swallow some very big "ifs." From two anecdotes we are asked to accept that hunting is "a common element in the backgrounds of most sexual predators." To make this statement, Clifton would, for starters, have to systematically study the records of convicted sexual predators to see if they are more likely to have hunted than the general population. To accept Clifton's view, one also has to overlook some confounding facts, not least of which is the fact that the number of men who hunt has been declining rather steadily for the past several decades (an issue I will revisit shortly) while rates of violent crimes, including sexual abuse of children, have been going up. With violence on the rise and the number of hunters declining, one has to imagine some very busy hunters out there to believe Clifton.**

Despite flawed statistical analysis and unwarranted assumptions, Clifton taps an ancient worry. Just as we legitimately worry about how reliably our celebrated athletes, from the high school football heroes to the superstar pros, can confine their aggression to the field or court (Guttmann 1986, 1978; Lefkowitz 1997), so too do we worry about how dependably those who hunt are able to restrict their appetite for killing to the field. The anxiety persists precisely because athletes do unleash aggression and hunters are ineluctably associated with the death of living things.

Think of the peculiar status those who deal with death occupy in society. The coroner and medical examiner do not enjoy the highest status within the medical profession. Morticians occupy a curious standing in the community that is not much relieved by being renamed "funeral director." We have moved our slaughterhouses farther away from our centers of activity and further from our consciousness, and even within the "meat process-

---

*Clifton is committing a classic statistical fallacy, known as the "ecological fallacy," by attributing to individuals the attributes of some aggregate of which they are members. It is like guilt by association. There are other statistical problems with Clifton's analysis as well. Adair (1995) has subjected Clifton's data to more sophisticated statistical techniques and discovered that the effects Clifton reports do not appear upon reanalysis.

**Violent nonsexual crimes began to suddenly drop in the mid-1990s for reasons the experts are still debating. No serious study of the rise and fall of violent crime so much as mentions hunting as a factor. See, for example, Blumstein and Wallman 2000.

ing" industry, clear lines are drawn between those who do the actual killing and those who ready various animal parts for our consumption.* Hunters enter a similarly charged symbolic arena in which their hands become bloodstained. No wonder hunters evoke decidedly mixed responses.

As he does in so many respects, Thoreau captures this ambivalence nicely. Thoreau hunted in his youth, and he clearly appreciates the way hunting intensifies a person's observational skills and understanding of nature. But he urges his readers to confine their hunting to their youth. "Such is oftenest the young man's introduction to the forest, and the most original part of himself. He goes thither at first as a hunter and fisher, until at last, if he has the seeds of a better life in him, he distinguishes his proper objects, as a poet or naturalist it may be, and leaves the gun and fishpole behind" (Thoreau 1992, 200).

Thoreau suggests what many, then and now, suspect about hunters— that they never fully grow up. Explicitly evoking the then popular theory that each of us, as we develop from embryo to senior citizen, recapitulates the developmental stages of the species, Thoreau reasoned that hunting (and fishing) were primitive arts once essential to our survival and a crucial "educational device" from which issued the first steps toward assembling a systematic body of knowledge about the natural world and how it worked. Hunting was, he implies, the origin of technology and science. But by Thoreau's day there was no longer the need for hunting—as Rush noted in 1790, agriculture had long since replaced the need for subsistence hunting, at least among the advanced civilizations, and science no longer depended upon observational skills honed by the desire to kill. All that remained was a developmental necessity, an experience from which to grow, to move on to more refined pursuits. Growing up is literally, in this sense, "evolving." Thoreau is torn between an admiration for the skills of the hunter and a revulsion at the coarseness that seems to go along with hunting.

Modern critics of hunting are far less inclined than Thoreau to see anything of value in the hunter. Even when they stop short of accusing hunters

---

*The French anthropologist Neolie Vialles has written a marvelous study of the history and structure of the French abattoir, *Animal to Edible* (1994), which shows how both the physical layout and the occupational hierarchy serve to circumscribe or compartmentalize killing, lest killing escape its ritualized containment. Though few societies have as clearly marked a hierarchy as the caste system of traditional India, where those who kill and directly handle dead animals are untouchables, slaughterers are usually near the bottom of the heap.

of being nascent serial killers or sexual perverts, the critics do not hesitate to impugn the intelligence, manners, self-control, or thoughtfulness of hunters. Writer Joy Williams (1990) blasts hunters for their killing ways. She writes: "For hunters, hunting is fun. Recreation is play. Hunting is recreation. Hunters kill for play, for entertainment. They kill for the thrill of it, to make an animal 'theirs'" (114). Ridiculing all attempts to contextualize the killing, Williams flays hunters for choosing to wallow in gore. Choosing thus, it follows that hunters must be low lives. "Hunters are piggy. They just can't seem to help it. They're overequipped . . . insatiable, malevolent and vain. They maim and mutilate and despoil. And for the most part, they're inept. Grossly inept" (121). Unwilling to concede anything of value in the character of those who hunt, Williams returns again and again to an insistence that hunting attracts the most base and ignorant among us and drags others who might be more worthy down to the same moral and behavioral level as that of the cretins whom she thinks predominate within the ranks of hunters.

Ron Baker (1985), leader of a group devoted to the abolition of sport hunting, offers a broader-gauged criticism of hunting than Williams. His opposition to hunting is primarily driven by his conviction that hunting for sport is cruel. But he is also alarmed by the way hunters have helped, directly and indirectly, perpetuate a generally exploitative relationship to the natural world. Hunters support a consumptive view of nature and have, Baker argues, used their political influence and their money (in the license fees and taxes referred to earlier) to tilt wildlife and habitat management activities decidedly toward game species to the disadvantage of almost every other living creature. Moreover, Baker argues that organizations representing hunters have more often supported extractive industries and the water projects of the Army Corps of Engineers than they have sided with environmental groups interested in preserving our natural heritage. When all of this is added up, Baker claims, the conclusion is inescapable: sport hunters are an environmental disaster. Instead of trying to bring hunters into environmental coalitions, as many of the "old line" environmental groups like National Audubon and the Wilderness Society occasionally do, Baker urges his fellow environmentalists to work to ban hunting.

Baker sees hunting as the cause of perverse environmentalism. End hunting, and somehow the money and energy that hunters have used to distort environmentalism will flow to real environmental causes. Curiously,

Baker has no explanation for why nonhunters have not been as generous with their time and money as have hunters. He would have us believe that hunters have blocked nonhunters from pursuing environmental initiatives that are not related to game fish and animals. This is, of course, silly. The plain fact is that until very recently, most Americans were unconcerned about wildlife and the environment. Hunters by and large had the field to themselves. While hunters and nonhunters certainly can clash over a particular issue (often the clash is over access: nonhunters are more inclined to save habitats by restricting access and declaring the place off limits to hunting), there are plenty of instances in which hunters and nonhunters have come together to advocate for protecting the environment. One very important example of this will have to suffice, lest we get bogged down.

In the mid-1990s a coalition of environmental groups, joined by groups who have long advocated for hunting as well as environmental issues (for example, the Izaak Walton League) started a campaign called Teaming With Wildlife (TWW); its goal was to pass legislation modeled on the Pittman-Robertson Act that would institute a tax on all outdoor equipment, from bird-watchers' binoculars to backpacks, the proceeds of which would be devoted to a range of environmental initiatives, prominent among them the enhancement of nongame species. Manufacturers whose products would be made a tad more expensive resisted (unlike the manufacturers of hunting and fishing gear), and grassroots support was too weak to move legislation along. Nonhunters, though vastly outnumbering hunters, were less a force for wildlife than hunters, not because hunters got in the way, but because too few nonhunters cared enough.*

Marti Kheel, an ecofeminist writer, goes well beyond Baker's critique of hunting in her essay "License to Kill: An Ecofeminist Critique of Hunter's Discourse" (1995). She argues that the thinking that has dominated American conservation/environmentalism is based on the same logic and ethi-

---

*The Teaming With Wildlife campaign morphed into a new initiative in 1999–2000, the Conservation and Reinvestment Act (CARA), which had unusually broad and bipartisan support in both the House and Senate; had it passed in full, it would have provided $900 million a year more or less, to a variety of environmental initiatives, including a significant amount dedicated to the enhancement of nongame species. Several prominent antihunting groups, including PETA, were conspicuously absent from the list of organizations urging passage of the act. Virtually every national group having anything to do with hunting and fishing signed on as supporters.

cal premises as sport hunting: nature is our plaything. She states her aim: "I hope to facilitate the divorce of environmental discourse from its blood-stained marriage to the activity of hunting, thereby exposing the true function of this discourse—namely, the legitimization of violence and biocide; in other words, the license to kill" (5). Kheel paints with a larger brush, but she shares Clifton's conviction that hunting is closely linked to all sorts of pathologies, especially those associated with violence and sexuality. But there is a crucial difference. Where Clifton is intent on showing that there is something wrong with those who are drawn to hunting, making them more likely than nonhunters to sexually molest children, among other unspeakable crimes, Kheel claims that the problem is men. Whether or not they hunt, men have been infected by their long association with hunting. Hunting is simply a symptom of a more general male aggressivity that yields, in other modalities, rape and murder. Kheel writes, "It is necessary to recognize that the perpetrators of violence throughout the world are, by and large, men, and the victims of this violence are primarily women and the natural world" (39).

Charles Bergman (1996), a professor of English literature and an animal advocate, refines and elaborates Kheel's argument, though he does not cite her essay. Using the tools of literary analysis, Bergman traces representations of the hunt through ancient myths and texts down to the present to show how hunting has been central to the way masculinity has been defined and is responsible for most if not all of the deplorable things commonly laid at the doorstep of men—war, serial killing, and, of course, sexual predation. For Bergman, as for Kheel, it is not enough to end hunting. Merely ending hunting would not, presumably, change the underlying psychosexual forces that work to produce aggressive males. Indeed, were hunting banned, men might direct all their aggression toward women. What Bergman and Kheel seek is some sweeping change in the psyches of men, such that they are no longer aggressive and violent. Ending hunting isn't enough: even the metaphor of hunting must be erased.

The connection Kheel and Bergman try to make between hunting and violence against women (or in Clifton's case, children) is irresistible for those inclined to put a stop to hunting. Cartmill (1993) is far more measured in his analysis of the cultural role hunting has played in human history, but even he can't hold back the urge to draw the analogy between hunting and rape. As he tries to fathom the motives of sport hunters, he wonders if at least a part of the attraction of hunting might be that it *is* bad, that evil is attractive, at least to some people.* "In fact . . . wrongness may itself be part of the attraction of hunting. Some men . . . seem to enjoy feeling evil; and some hunters—say, those who kill wild animals for no discernible reason . . . —may enjoy their sport precisely because it makes them feel wild and wicked and crazy. . . . [It is] the rural equivalent of running through Central Park at night, raping and murdering random New Yorkers" (239).

Several paragraphs later, Cartmill sums up: "Perhaps the real social pathology linked to hunting is not war . . . but rape" (240). But why, we need to ask, ought we seek a link between hunting and social pathology? The answer brings me back to where I began this brief examination of the worry hunting gives rise to: killing. More pointedly, the further removed hunting gets from being necessary to one's own survival, the more it becomes a matter of choice, the more those who hunt have, like it or not, a desire to kill.

Of course, directly or indirectly, each of us lives at the expense of many other creatures. Even the most conscientious vegan is complicitous in the demise of animals who are displaced by the farms that grow legumes, rice, and green vegetables. Animal products enter our lives in countless ways such that the elimination of all dependence on animal death is virtually impossible. But Cartmill again puts his finger on the key concern. "[E]njoying the fruits of the kill is not the same thing as *taking pleasure in killing itself.* Even the most enthusiastic lover of fried chicken may suspect that there is something wrong with a man who finds recreation in wringing the necks of pullets" (241; emphasis added). There it is—a person who finds pleasure, indeed, deep satisfaction, in killing a living thing must be different, somehow less decent or refined and, conceivably more dangerous, than those who avert their eyes or otherwise ignore where their food, clothing, and shelter come from. Clearly, we need to know more about who hunts.

---

*Random rape and murder aside, Cartmill has a point to which I return in a later chapter—going into the "wild," whether to backpack, fish, hunt, or to party, entails leaving the constraints of society behind. In the woods, the burden of following rules is yours alone, for the chances of being found out for misbehavior are slight to vanishing. Some wear that burden more lightly than others. I shall explore this tension between going out to "be free" and going armed, not only with a weapon, but with a self-imposed code of behavior.

II

WHO HUNTS?

Hunters have always been in the minority, even in societies where hunting was one of the main sources of food. When hunting ceased being a necessity, the ranks of hunters no doubt thinned, if for no other reason than the shift to horticulture required more hands to labor. The greater predictability of crops made hunting increasingly marginal, from a utilitarian point of view, and meant that only those with time on their hands could hunt. In general, this meant the young (before the duties of supporting a family began), the poor, and the well-to-do were the likely hunters. As in so many other aspects of our cultural history, we know most about the hunting of the well-to-do, whose days afield often resembled a military campaign and resulted in staggering amounts and varieties of birds and mammals killed (Blüchel 1997). Much less is known about how the poor hunted, no doubt because they did so furtively, and, even when they were not trespassing or poaching, they did so without fanfare. Aside from efforts to control predators, nonelite hunters were generally regarded as idlers, wastrels, and brutes, scarcely less wild than the creatures they hunted. In the United States, as I argued earlier, this low reputation began to change in the middle of the nineteenth century.

By the early twentieth century, hunting was thoroughly folded into a national narrative that celebrated the taming of the continent and the

41

dawning of an age when engagements with nature could be structured by a sporting ethic which borrowed elements from elite traditions but which was thoroughly bourgeois: the emphasis was on self-restraint, regard for property, refinement of technique, and the embrace of objective natural history. There were still plenty of hunters drawn from the margins of society, both the working class and the rural poor, but they were rapidly becoming incorporated into middle-class culture, if not middle-class living standards (Warren 1997; Herman 2001).

Magazines specifically devoted to hunters and anglers, and columns devoted to hunting and fishing themes in daily newspapers, regularly promoted the ideal of the sportsman, a person seeking the satisfaction of having hunted or fished well, defined in terms of fair chase and mastery of technique, not the quantity of fish or game reduced to possession. Thus firmly linked to the moderation that lay at the heart of the conservation ideal, hunting began to become reputable. Hunters joined the yeoman farmer as protean Jeffersonian democrats. It remained the case that only a minority hunted, but the minority was portrayed as and very probably felt that they were embodying values endorsed by the entire society. An ideological base was thus laid. When, after the end of World War II the middle class swelled, suburbs multiplied, and trade unions won paid vacations, rising wages, and a measure of job security for much of the nation's blue-collar workers, Americans in rapidly expanding numbers had both money and leisure time to pursue hobbies of all sorts; visits to state and national parks soared, the ranks of bird-watchers grew, and, of course, fishing and hunting grew in popularity, with hunting, as judged by license sales, peaking in the late 1950s.

The popularity of most of these outdoor activities continues to be robust, but beginning in the 1970s, participation in hunting stagnated and, by the 1980s, began a slow but steady decline. Various studies (Heberlein and Thomson 1996; Duda et al. 1998) all reach the same conclusion: the percentage of males engaging in hunting has been steadily eroding while the participation rates of women have risen sharply. Heberlein and Thomson, for example, report that between 1980 and 1990, male participation rates fell from 19.5 percent to 16.4 percent (85). In 1980, 1.5 percent of women hunted; by 1990, 2.7 percent of women were hunting (Heberlein and Thomson 1996). This rate of increase is large, but since the percentage of women hunting is still low, the net result is that after decades of slow but steadily declining percentages of hunters, the ranks of the nation's hunters have stabilized at just under 10 percent of the sixteen-year-old and over population (Duda et al. 1998). Apart from the increasing presence of women, the most notable feature of these trends is the sharp decline in the recruitment of young males to hunting. I will revisit this decline in the concluding chapter when I turn my attention to the future prospects for hunting. For now, it suffices to say that with the stream of young males into hunting on its way to becoming a trickle, the average age of the nation's hunters is steadily increasing, and now hovers around forty-five years of age.

Hunters are still predominantly male, of course. In the National Opinion Research Center 1998 General Social Survey, 80 percent of the respondents who hunted were male. Hunters are even more overwhelmingly white: 91 percent of the hunters are white, compared with 77 percent of the nonhunting population. Nearly two-thirds of hunters are Protestants, compared with just over 50 percent of nonhunters. Half are married and a quarter are separated or divorced, both slightly higher than the comparable figures for nonhunters. Nonhunters are more likely to have graduated from college or to have earned graduate degrees (26 percent) than hunters (16 percent), though there are slightly fewer hunters than nonhunters who have not completed high school, no doubt reflecting the absence of recent immigrants and minorities among the ranks of hunters. This difference in educational attainment, though, does not translate into lower incomes for hunters, no doubt because, as we have just seen, hunters are overwhelmingly white and male: in our society, almost any group with a preponderance of white males will compare favorably in terms of income with a more diverse group, even if the more diverse group is somewhat better educated. When we look at a measure of overall social status that takes into account education, income, and occupational status, however, we find that hunters have a somewhat lower median score (469) than do nonhunters (495). This suggests that the "typical hunter" is more likely to be found in the lower-middle and working classes than in the middle and upper-middle classes, but the differences are not huge.

The sharpest demographic difference between hunters and non-hunters is their place of residence: half of all hunters live in the nation's smaller cities, towns, and rural areas (compared with only one-quarter of nonhunters). Though tilted toward small cities and towns, hunters are widely if not evenly distributed across the nation, with rates of participation highest between the Alleghenies and the Rockies and lowest on both coasts. Other than the fact that hunters are disproportionately white males who live outside of the nation's urban centers, there is nothing in this demographic profile that would call attention to hunters. Their occupations are somewhat humbler and their incomes a tad higher than the general population's. On the face of it, there is nothing in these data that would raise suspicion or that would suggest that hunters are the scourge that the critics of hunting allege. Though data that would allow me to directly examine the critics' allegations do not, so far as I know, exist, NORC's General Social Survey includes questions that do allow us to compare hunters and nonhunters on a wide range of attitudes and behavioral patterns. Though this information is by no means perfect, with it I can go well beyond the demographic profile I have just summarized. Before I proceed, however, one general observation needs to be recorded. Just as with income, many of the differences between hunters and nonhunters turn out really to be differences between males, whether or not they hunt, and females. Hunters are, as noted, preponderantly male, whereas the nonhunting public is more nearly evenly split between males and females. So, for example, hunters are much more likely to report that they drink alcohol than nonhunters, but the difference is sharply reduced when we compare male hunters and male nonhunters. Both groups of men are more likely to drink than either female hunters or female nonhunters. What follows is a brief summary of the more detailed data contained in the appendix.

The most remarkable thing that emerges from a comparison of hunters and nonhunters is how unremarkable the differences generally are. So, for example, in the battery of nine questions regarding attitudes toward the environment, the only statistically significant difference that emerges—and it is a very slight difference at that—is that male hunters are a bit less inclined to bind American environmental policy to international agreements. The charge that hunters do not really care about the environment, except insofar as it be maintained to give them hunting opportunities, is not supported by these data. This is all the more striking because the only clear pattern of difference between hunters and nonhunters emerges around politics and political ideology: hunters, especially male hunters, are more politically conservative. In 1996, Bob Dole got nearly 50 percent of the male hunters' votes while only one-third of male nonhunters said they voted for him. Almost half of the male hunters say they are conservatives, compared with a third of male nonhunters. But when it comes to the environment, hunters appear to shelve their conservatism since they are as

supportive as nonhunters of the federal government's regulating both individuals and businesses to protect the environment.

It turns out, not surprisingly, that more hunters, again especially male
hunters, than nonhunters are drawn to what nowadays are called "traditional values." They believe that people should stand on their own two feet
and be held responsible for their misdeeds. Male hunters are much less
likely than their nonhunting counterparts to approve of unrestricted access
to abortion and considerably more likely to disapprove of homosexuality
(72 percent versus 57 percent said that homosexuality is "always wrong").
Even though hunters are more conservative on a range of social issues,
overall the differences, even when statistically significant, are not huge.
These data, at least, provide no support for the notion that hunters are far
outside the mainstream of American politics.

It also appears that the conduct of hunters and nonhunters in the private
realm is remarkably similar. Patterns of church attendance are virtually
identical for the two groups. Hunters are drawn to X-rated movies and sexually explicit websites to the same degree as nonhunters and are slightly less
likely to have paid for sex than nonhunters. Hunters and nonhunters are
equally likely to have been faithful while wed. The only difference of note,
really, and again it is not large, is that hunters, especially female hunters, are
more sexually active than nonhunters—they report having had sex somewhat more frequently and are much less likely to have been abstinent.
Hunters, as I have already noted, are somewhat more likely to consume alcohol, and more hunters than nonhunters admit that they sometimes drink
more than they should. The differences, however, are not large and they
are not statistically significant.

In short, there is nothing in these data that gives much ammunition to
those critics of hunting who insist that hunters are somehow deviant and
inclined to perversion. Of course, I have no way of knowing how many
hunters are pedophiles or rapists or abusive of their spouses—information on these sorts of things are not asked in the General Social Survey
(and even if such information was sought, it is not clear how reliable the
responses would be). By the same token, I have no way to compare
hunters and nonhunters in terms of criminal records. All I can say is that
the data that is available would not, of itself, give rise to suspicions about
the character and behavior of hunters. Indeed, given the scant differences between hunters and nonhunters that show up in the General Social Survey, one might reasonably wonder why hunters have attracted so

many detractors. This is a question I address in the concluding chapters, after I have explored hunting, with the help of the men and women hunters I interviewed.

Statistical profiles are invaluable for giving us an idea of the broad contours of a population and allowing us to compare one population with another. Statistics are less useful for conveying the shape of character, the depth of feeling, or the particularity of experience that forms a person. For these things, we need to understand people in the context of their lives. That is why I set out to interview a randomly selected small sample of hunters. My intentions were not to replicate the sort of survey that NORC mounts but, instead, to interview hunters with an eye to capturing their unique experiences and perspectives, the sorts of information that cannot be captured by statistics. My goal was to gain insight into why people hunt—what it means to them—as well as to understand how they think about killing animals and how they see themselves in relation to the larger culture, a culture that is increasingly critical of hunting. Even though I will quickly leave statistics behind, it will be helpful to have a modest statistical profile of the people I interviewed. Sixteen percent of the thirty-seven people I interviewed were women, roughly the same percentage as found in the NORC survey. Eleven percent were under twenty-five years of age, 54 percent were between twenty-five and forty-five years of age, and 35 percent were forty-five or older. Skilled workers predominated (51 percent), with unskilled workers adding another 19 percent, giving the group a decidedly blue-collar character. Sixteen percent were white-collar workers and another 14 percent were professionals. Few were poor (11 percent made less than $20,000 and they were young people still living at home), 40 percent earned between $20,000 and $50,000, 38 percent earned between $50,000 and $80,000, and 11 percent earned over $80,000. Though I cannot say with certainty that these characteristics make these people representative of all hunters in Massachusetts, they are certainly not wildly out of character with the profile derived from the General Social Survey.

Over the course of the several chapters that follow, I will introduce each of the men and women I interviewed. Each was unique and had interesting and important things to contribute to an understanding of hunting and what it means to hunt. Before I begin my detailed examination of what it means to hunt, I would like to offer several brief profiles of people who represent the range of experience and perspectives that I encountered. I will start with the Osgoods.* I initially interviewed Sioban, a nineteen-year-old girl, and was so struck by what she said about her family, that I made another trip to interview Sioban's father and mother.

Carl, his wife Bev, and the youngest of their two daughters, Sioban, live well off the beaten path in rural central Massachusetts. Sara, the oldest daughter, lives in a town forty-five minutes away, near her job with a rapidly growing aquaculture enterprise, her first job after graduating from the University of Massachusetts with a degree in environmental science. The Osgoods' nearest neighbor is roughly a half mile away and things will stay that way because the Osgoods' house and thirty-five acres are surrounded by land owned by Massachusetts Audubon and maintained as a wildlife sanctuary. Carl and Bev chose this site because they wanted a place that was not going to get built up. They wanted to look at trees and meadow, not subdivision lots. Though they have a large vegetable garden and a freezer well stocked with the fruits of the hunt, the Osgoods are not back-to-the-earth types or recluses. Carl is a college graduate who decided he preferred the freedom of a carpenter/solo contractor to the nine-to-five of the white-collar world. Bev attended but did not complete college and, now that her daughters are grown, works as a rural route mail carrier. While not rolling in money, they hunt and garden for pleasure, for the satisfaction it gives them, not because they cannot afford to buy food. The Osgoods take pleasure in feeling intimately close to nature. Their open and airy house is filled with wildlife photographs and mounts of a wide variety of animals they have shot or trapped. Whether indoors or out, the Osgoods are immersed in nature.

The Osgoods' house is modest, as they themselves are. They have blended work, gardening, fishing, and hunting into a nearly seamless round of activities that comes about as close to self-sufficiency as one can and still remain firmly linked to the larger society. The outdoors is their entertainment as well as a source of nutrition. When they are not hunting, they are scouting and exploring, keeping track of the comings and goings of the abundant wildlife around them. When I first visited the household to interview Sioban, the table at which we sat was covered with binoculars and cameras and a variety of lenses. It was summer, and when family members had some free time, they grabbed what they needed from the table and headed out to observe and record the natural world in which they were submerged.

---

*All names have been changed to protect the identity of the men and women I interviewed.

Perhaps it was this amateur naturalism that drew the Osgoods to keeping the local taxidermist in steady work. Their house was as near to a diorama of the wildlife of the Northeast as any dwelling I will ever see. Though I am not a fan of taxidermy, I found the Osgoods' display curiously tasteful—it fit them well. These were not monuments to vanity; they were tributes to nature and to the enterprise that is required if we are to live on the land. The Osgoods are not looking for bragging rights. They are bringing the creatures they revere into their lives in yet another way. Hunting, for them, as it is for many, is one of the ways to merge with nature, to be an active participant in the biological life cycle. They aren't so much dominionistic as reverential, to borrow terms from Stephen Kellert (1996). This reverence is directed at animals, but it springs from something more encompassing: a reverence for a way of life the vast majority of us know only through history books or, more likely, from Hollywood representations of our agrarian past. Even with all the links to the contemporary world—college education for children, power tools, automobiles—the Osgoods still manage a fuller engagement with the past than most of us imagine possible. Their gardening and hunting also keeps them close to one another, and through the sharing of game and garden produce, they are linked to a network of friends and relatives.

Sioban was home for the summer when I interviewed her. She had just finished her first year at one of the region's many small liberal arts college and was working as a lifeguard (she was on the swim team in high school and college) and helping her father on jobs that required an extra hand. Sioban, bright and attractive, could easily pass for a teenager preoccupied with getting a tan and being seen in the right crowd. But, all in all, she'd just as soon be in the woods. Sioban could not remember when she first got interested in hunting. "I never thought about it," she told me. "Hunting's always been there." Both she and her older sister grew up with hunting. Both are very capable—there are many hunters twice their age and older who have gotten fewer deer than these two young women have. But I got no sense that this sort of comparison mattered in the least to her. There was no competitive edge to the way she spoke of hunting. When Sioban spoke of the last deer she got, a doe, she was not in the least boastful, though she did admit that she took pride in having mastered the woodcraft and the proficiency with a shotgun that killing a deer requires. Hunting has taught her a lot.

For a long time, her dad was her instructor, and he and Bev were clearly pleased with how each of their daughters absorbed the lessons—the lore, skill with compass and maps, knowledge of the habits of animals, and, of course, gun safety and marksmanship. But Sioban, at some point, began to learn on her own, to go beyond applying the lessons her father had imparted and to make her own decisions. Hunting also taught her things about herself—how much discomfort she could endure; the strength of her resolve to behave responsibly in a context in which only she was judge; her ability to think clearly and calmly when excited or stressed. Knowing such things has given Sioban a strong sense of self. She is poised and self-confident because she knows she is competent and capable. This came across at many points in the interview, but one comment stands out. We were talking about the way hunting tests one's ability to live by one's wits. Sioban said, "I'm really proud of the fact that I'd know how to handle myself if I got lost in the woods. I've paid enough attention to my father and my grandfather and grandmother (we're all outdoorsy people) so that I know what berries and things I can eat and I know how to catch a fish or an animal for food. It makes me feel good about myself to know that I can survive."

Surviving in the woods is comparatively simple compared to surviving in a society in which women are not expected to be "outdoorsy." She and her father both had stories to tell about men who thought girls should stay home and play with dolls. But that seems never to have been a question for the Osgoods. Bev recalled that when she was pregnant with Sioban's sister, friends gave her a baby shower. One of the gifts was a toy rifle. Still, there is resistance to be dealt with. Sioban told me how odd the people she had just met at college thought she was when she went home on fall weekends rather than stay and revel in their newfound freedom. Her college friends thought it was "cool" that she had worked alongside her dad as a carpenter, roofing, framing, and the like, but they were scandalized when they learned that she was going home to hunt. "They gave me a hard time," she said, savoring the irony that her classmates thought they were so advanced in their rejection of male/female stereotypes. I asked her if that sort of disapproval made her feel awkward or uncomfortable. "Not really, because I'm kinda confident. I like to shock people like that. I love it when we're in camo and we go in to get a soda or something after we've been hunting. I take my hat off and my hair comes down and everyone goes "whoa." I love that. It doesn't make me feel awkward. I love to surprise people and just say 'Hey, I can do it too, you know.'"

I do not know how self-consciously Carl and Bev set out to create a situation in which they could raise children to be competent and confident in two very different worlds—the contemporary world and a world that more nearly resembles late-nineteenth or early-twentieth-century rural America. But that is what they seem to have done. Hunting has been a key factor in fashioning this accomplishment. Hunting was the context in which Carl and Bev could impart values to children that in almost any other context would seem either hopelessly quaint or so detached from the real world as to seem little more than a string of clichés. In the context of teaching a youngster to hunt, however, the lessons of responsibility, sharing, humility, and consideration for others, including animals, make sense because they are concretely applicable and they make sense where it ultimately counts most—on the ground. It's worth quoting Sioban at length in this regard.

> My friends from school will ask me "What are you doing this weekend?" And I'll say I'm just going to hang out with my parents. They don't understand that because they don't live that type of life, they don't really respect their parents. But my parents are everything to me. I think that a lot of children today are missing that respect. I respect my parents, and that taught me to respect others too. As a lifeguard I see a lot of kids at the beach who have no respect for anything, not even themselves. I think a lot of people really miss out when they don't have anything they share with their kids or kids with their parents. The kids don't really know them [their parents] and they [the parents] don't really know their kids.

This was not offered in a preachy or moralizing manner. It was more a lament, a sad commentary on the way things are. The Osgoods have carved out a niche for themselves where it appears they can in some sense have the best of both worlds. Their daughters can get good liberal arts educations that give them access to decent jobs and the broader culture and at the same time be nourished by a deep engagement with the family and with the out-of-doors. "We don't have to lock our doors out here," Sioban remarked, as if to summarize her good fortune.

To be sure, life hasn't always been easy. Carpenters do not always have steady work, and in New England work slacks off in the winter even in the best of times. The garden and game, although not a necessity, are certainly a reassurance. In Bev's words: "I think it goes back to that survival instinct. If for some reason you couldn't work and couldn't buy meat for the table, you know in your own heart you can provide for your family. And there are a lot of people in this town, let alone in the state, who don't have that inner feeling. It is a peaceful feeling." Carl agreed but was quick to note that the satisfaction was real, as a feeling, but it was not possible to really "go primitive," as he put it. If there was some economic calamity that forced people to fend for themselves, Carl said with a wry smile, "We (Americans) are into a lifestyle now where if we got into that situation, we'd have to eat other people, because there's not enough game out there, but there's a whole lot of people."

Before Bev began working full time, the garden was larger and was worked as much for augmenting Carl's income as for good feelings. In those days, game also helped the family stretch their dollars. But now, with Sara on her own and Bev working full time, hunting and gardening have become purely recreational. But Carl, Bev and Sioban, each in his or her own way, took pains to be sure I understood what they meant by "recreation." Bev left no ambiguity: "But they (the girls) were taught from the very beginning that yes, it's fun to be out in the woods, and yes, the hunt is fun, and yes, it's nice to kill a deer, but there's still *responsibility* (her emphasis) and there's work. Dad isn't always going to be here to dress the deer for you. Responsibility goes with the pleasure." To illustrate the point, Carl related this anecdote: "My buddy from northern New Hampshire came down to hunt with us last year. He just couldn't believe it. Sara shot a five-pointer, and when we finally got to her she had it all gutted out. She was blood from head to toe. This guy went nuts. He says, 'I can't believe this.'"

On another occasion, Sioban had shot a deer late in the day and was due back at college that night, a long drive from the place they were hunting in New Hampshire. She was torn between cleaning up and getting back to school or finishing the dressing of her deer and returning to campus looking more like an ax murderer than a coed. Carl, too, was torn. He'd brought his girls up to take responsibilities seriously, and there was no more serious a responsibility than that owed an animal you have killed. Allowances were made, Sioban showered before heading back to school, and Carl finished caring for the meat.

It is precisely in moments such as these that the moral seriousness of the hunt gets established. Critics of hunting often think that because the meat is not essential, because people hunt for "sport" or "recreation," they are having fun at an animal's expense. Of course, hunting is fun, as Bev

Osgood acknowledges without hesitation. But it is a particular kind of fun. It's fun that carries a great and humbling set of obligations, at least if you want the respect of people like the Osgoods. Not everyone holds themselves to the Osgoods' high standards or has managed to sustain such a close family, knitted together by a mutual love of wildlife and a delight in the hunt. For the members of this family, hunting forms a bond among them and between them and the natural world. It is an activity that teaches large and small lessons, the import of which extends far beyond the woods and lasts long after the echoes of the season's last shot have faded. Hunting sustains them and it sustains a sense of continuity, a sense of rootedness, from which the young daughters derive strength even as they come to understand just how tiny we all are in the presence of the wild. Hunting may be fun, but for the Osgoods, the fun is very serious, even worshipful.

A few days after my second visit to the Osgood household, I met Bill Crafter at his home in a small town in northwestern Massachusetts. Bill Crafter is a young seventy-year-old, retired physical education teacher and high school coach who, in recent years, has hunted no more than two or three times a season, even though he now has plenty of time. He hunts mostly to share time with the one of his four sons who lives nearby. Bill grew up in rural Massachusetts and hunted avidly as a teenager and for a year or two after he got out of the Navy. But his own education got in the way of hunting, and then, when he began teaching and coaching, there just wasn't much time—afternoons were spent practicing with the team and Saturdays were game days. (Hunting is not allowed on Sundays in Massachusetts, the last of the blue laws to survive the tide of secularism.)

All of his sons were introduced to hunting, though not by Bill. Both Bill and his wife came from hunting families, and they have relatives in Vermont and Massachusetts who are avid hunters and were only too happy to introduce their nephews/cousins to hunting. Of the four, only one still hunts. The others, a medical researcher, a minister, and a high school teacher, are too busy with jobs and raising families to have the time or inclination to hunt. The son with whom Bill hunts occasionally is an administrator at a prestigious prep school, and though his is a demanding job, he has enough flexibility to be able to hunt locally with some regularity in any of three states—Massachusetts, Vermont, and New Hampshire; all offer good hunting within the radius of an hour's drive or less from the small town in which Bill and his son both live.

"So what do you do with your time," I asked. "My wife and I walk a lot and we love to garden," he said, and then, almost off-handedly, he added, "oh, and I carve birds." Indeed he does. My eye had been caught by a beautiful carving of a loon resting over the fireplace in the den where we were talking. It was one of his carvings. He appreciated my interest, but when I expressed admiration he brushed it aside. "That was one of my first efforts. Here's a much better piece of work," he said as he handed me a carving of a pintail duck. It was as close to a living thing as a block of wood can be made to be. He estimated it had taken him four or five hundred hours to complete.

He was especially fond of pintails. "I got interested in them because I used to hunt ducks and I admired them—the way they could take off from the water and go straight up." Now, he couldn't imagine shooting a duck. In a sense, though, he still does hunt ducks. He scouts the shore of the Connecticut River, a short drive from his home, looking for ducks. When he spots some, he gets as close to them as he dares and studies them through binoculars, watching how they move, how they rest in the water, and, of course, the shape and color of their feathers. And then he captures them in wood.

Bill is one of several of the hunters I interviewed who is no longer interested in killing. For some, this was something that gradually became a conscious decision to shoot less and less often. For others, like Bill, there had never been much of a desire to kill. Bill has shot ducks and most other game birds, and he shot one deer, years ago (he stopped deer hunting in the early 1950s because he did not feel safe in the woods with so many hunters chasing so few deer); but someone got to the animal before he did, and the other guy claimed he had shot the deer. Now, he will shoot a pheasant if his son's bird dog has worked it well because he wants to honor the dog's work and his son's training. Hunting, for Bill, is more a ritual or a rite than anything else. Bill thinks that hunting and firearms are connected in a long skein of events that reach back to the roots of our nation's independence and to the hardy individualism for which Americans are known, precisely the national narrative that I discussed in chapter 1.

Bill's reasoning was straightforward: the fact that a lot of people hunted meant that there were lots of people who had a gun and knew how to use it. Years ago, Bill offered a physical education course devoted to skeet shooting as part of the high school physical education program. Students had to bring their own shotguns; the school provided the ammunition. Some people were alarmed to see kids carrying guns to school, but they got over it when it became clear that the program was popular and well run. There were no accidents, and students clearly learned important lessons in safe handling of shotguns as well as acquired competence. We both reflected on why it would be impossible, today, to have such a program. I then broached the subject of gun control. Bill was firm.

No, I'm against that. I just feel that as an American citizen we have the right [to own guns] and it's built into the Constitution, the Bill of Rights, and from the history of our Revolution. Where would we have been if we didn't have guns? We wouldn't be a free country today. I'm not saying, like some do, that we need guns nowadays to preserve our freedom. I'm not much worried about that. I don't agree with the NRA [National Rifle Association] any more, though I once supported it and even belonged to it for a while. But I believe strongly in the right to bear arms, the right to have guns in citizens' hands. I think it's a traditional part of our history and important to the country.

Bill was, like many of the men and women I interviewed, politically liberal and culturally conservative. Actually, Bill preferred the word "traditionalist" to "conservative" because the latter word carried political implications with which he did not wish to be associated. Hunting was one of the ways Bill renewed and reaffirmed his traditionalism. Getting out the gun, working the action to make sure everything is right, pulling on the boots—all the myriad things that people have been doing for a long time in preparation to go forth in search of game he does. In this, even though he is scarcely a hunter by comparison to the Osgoods and even though he has no interest in killing a pheasant, Bill shares a great deal with those who are more avid, more bent on getting game. Though he acknowledged that he looks forward to "a good venison steak once in a while," compliments of his relatives in Vermont, he is content "just getting out to inhale the crisp air." In a very real sense, he is affirming his citizenship, paying homage to those who made this country what it is.

At the other end of the state, in a neighborhood of trim older two- and three-story houses just a few blocks away from the rail line that carries commuters to and from Boston each day, I met another retired educator. Like Bill, Peter Hanks had to largely suspend his hunting and curtail his fishing when he entered the classroom as a young man out of college. He had started hunting when still a youngster—his father got him his first gun when he was eight or nine. Though he went hunting occasionally with his father, from whom he learned the elemental rules of gun safety, most of his hunting was done with the husband of his oldest sister. They went small game hunting and deer hunting together until Peter was a young adult. But once he began teaching, there was just too little time since he lived in the most densely settled part of the state and would have to drive considerable distances to hunt. Fishing was easier because he had summers "off" (there was plenty to do to prepare for the next year but at least his schedule was flexible). Then he became a principal, and even his summers got encumbered.

Retirement agrees with Peter. He is short, trim, and when I visited in mid-summer, he was well tanned. The reason was simple: "all I do now is fish and hunt," he said as we settled into comfortable chairs in the dining room that had been one of the additions he and his wife, also a retired schoolteacher, had made to their Cape as their family had grown. As I was setting up my tape recorder, his wife came in from walking their beagle, a dog mostly known for rabbit-hunting prowess. "Rabbits are zilch around here," Peter said a bit ruefully, "but she's a good pheasant dog. Put up [flushed] nineteen last season." He could not be more pleased with how things were turning out for him personally. He was healthy and had both time and money enough to take full advantage of the abundant game New England offers. Last year he hunted grouse and pheasant through the fall, deer in December, rabbits through the late winter, and turkey in the spring. Now he was fishing—he especially likes to fish one of the state's prettiest stretches of trout water, the Upper Deerfield, which is restricted to fly fishing and catch-and-release. We talked a bit about the apparent irony of a hunter liking catch-and-release. Peter didn't feel any contradiction because he didn't like to eat fish. Like almost all hunters, Peter believed that eating what you kill is obligatory. Killing a fish and not eating it would be wrong so catch-and-release was just fine with him.

Though he was thrilled to be getting back into hunting, his enthusiasm hadn't overwhelmed his judgment: he was well aware of his limits, of how much he had to relearn or learn for the first time. Peter had gone turkey hunting for the first time in his life (when he was hunting as a young man, there were no turkey to hunt in New England), and the experience was still vivid. Hunting turkey in the spring, when only males are legal, involves stealth—the hunter wears head-to-toe camouflage and with one or another calling device fashioned from wood, slate, or rubber tries to lure a tom into close range by making the sounds of a hen looking for a mate. Because the tom is a large bird and its feathers are like armor, the hunter has to coax the wary bird into quite close range in order to insure a clean kill. (A wounded bird will almost never be found because turkeys are capable of running long distances at amazing speed.) It sounds easy but it is not. The tom is approaching to mate, and its feathers are puffed out and its wings are semispread, making the bird look half again as large as it really is. There is high drama in this, and it is hard to maintain composure. Here is Bill's account of what happened on his first outing:

> I went to the same area I had hunted deer in last fall, but this time, I went alone. To go into the woods alone early in the morning, well before light, gave me a kind of eerie feeling, and I couldn't help thinking "this is crazy." But then I realized that the quiet was special, and as the sun began to come up and the woods began to come alive I really enjoyed that. I saw animals and birds up close like never before. Then I called for a turkey and one answered. After that, I think I broke every rule in the book. I got him to come in to about fifty-five yards. I talked to that bird and the bird talked back to me. That was the biggest kick that I've had in a long time. I'm saying "this is easy." I was calculating the range and had a spot picked out where I knew if he got there, I could make a good shot. I could see him clearly, but all of a sudden he stopped approaching me. I must have blinked or something. The next thing I knew he turned sideways, gave one look in my direction, and vanished. Easy my foot.

The rules Peter said he broke were the "rules" of the tactician, not the game rules or the rules of safety. As the account makes clear, he followed those rules faithfully. A more impulsive person, a person Aldo Leopold would describe as having "trigger itch," would have blazed away at inappropriate range. He explained his restraint in these terms: "I'm not out there to make an animal suffer. If I can make a clean kill, then that's exactly what I'm going to do, and I practice so that I can do that. After all, I don't need to kill that bird." If not to kill, why did Peter go hunting?

Peter's answer, echoed in one way or another by everyone I spoke with, combined a desire to get close to nature and at the same time to test oneself. The testing involved, in different intensities with different people, endurance, woodcraft, tracking, matching wits with quarry, and, of course, proficiency with a gun or bow. For many, the tests also involved a curiosity about how one would have fared when America was a rural society. Here's how Peter captured these elements:

> It's being in the woods and then becoming part of what's happening right there. To have things moving around you without being aware that you're there, to be able to see things that you'd never dream of—a fox carrying a rabbit back to its den, a woodchuck doing something you didn't know they could do. You come across things that are amazing and you feel part of the landscape. And then once in a while I say to myself if I were living in a different time, would I be someone who worked in a field, or would I be a hunter? Could I depend on being able to fool trout or get deer to come within range? I haven't been that good because I'm just getting back into this, so now I think when I come across a deer we've just bumped into each other. It's not my ability to stalk a deer now. But it's part of what I'm looking forward to, becoming more adept in the woods.

Peter was more than a little worried about hunting's future. Having been away from hunting for so long, he was perhaps struck more vividly by the loss of hunting coverts, particularly of places he remembered from his youth, than someone who experienced the change gradually over a span of years. Much of the land had been built up, and a lot of what remained open had become posted. It bothered him no end, not for selfish reasons so much as for the cultural shifts these changes represented. On the one hand, he was sure that we have been far too reckless and shortsighted in the ways we have used the land. On the other hand, he was also sure that much of the opposition to development and to hunting was based on false ideas about both nature and hunting. Either way, nature loses, whether at the hands of exploiters or of those who think they are protecting it.

Having spent most of his adult life in the classroom and as an elementary school principal, Peter was well positioned to observe how nature gets depicted in school curricula. Especially in the younger grades, he said, animals are depicted in almost cartoonlike ways and discussed as though they had human qualities. Worse, "they have big eyes and the human qualities they have are only the good ones; they never have any bad human qualities." He continued, "Everything in nature is right. It doesn't rain in this part of the forest, animals don't starve during the winter, and there's no fox or owl tearing apart a rabbit. No wonder kids grow up with romanticized ideas about wild animals and nature. No wonder people are unaware of what wildlife and the woodlands and fields are really like. It's not a 'Bambi world.'"

With fanciful ideas about animals and nature so widespread, he was not surprised that hunters have come to be held in such low regard. Few parents of the children in his school—or anyone else for that matter—knew that the soft-spoken, gentle man in whose care they entrusted their kids was a hunter. "They knew I fished, but not that I hunted. I guess it's my dark side," he said, smiling impishly when he said the words "dark side." He went on to tell me that when his neighbor of many years recently learned that he hunts, she said to him, "I can't imagine you hunting, you are such a nice person."

Nice people don't hunt. We laughed at the absurdity of that sentiment, though the laughter had an edge to it. It was an absurdity that was broadly subscribed to. Peter is, his love of hunting notwithstanding, a nice person. He also belongs to the NRA and is an alternate on his rod and gun club's pistol team. He's not wild about the NRA's politics or their growing intransigence to any and all efforts to control access to exceptionally lethal weapons, he has no problem with tightened registration and licensing regulations, but he noted that the NRA is the only organization protecting what he believes is his right, as a lawful, indeed even a "nice," person to own firearms. Like Bill Crafter, Peter is somewhere on the moderate to liberal end of the political spectrum and has no sympathy for the right-wing agenda of many supporters of the NRA. In a sense, he feels trapped into supporting the NRA. But he is determined to see guns and hunting persist.

Peter's only grandchild is a girl, not yet a year old when I interviewed him, but he was already planning their fishing trips together, and he had two small-bore shotguns stored away, one of which would be hers when the time came. "We'll see what happens from there," he said pensively. If present trends continue, Peter's granddaughter, if she takes to hunting, will have plenty of women with whom to hunt. Two of them might be Jim Ramada's rambunctious daughters.

Jim arrived late for the evening interview we had scheduled. It had been a glorious early summer day: clear, dry and warm. It was the kind of day that vacationers and fly fishermen dream of. It was also a perfect day for house painters. Jim would have preferred being a fly fisherman that day, but he paints houses for a living. He apologized for being late, but he had taken advantage of the good weather and worked longer than usual. While I was waiting for him to arrive, I studied his modest house for any sign of hunting—deer antlers over the garage door, a hay bale bow target in the back yard, decals on the garage windows, or a dog run beside the house. There were such signs evident on other houses in the neighborhood, as well as several boats clearly rigged up for fishing rather than for water skiing or pleasure cruising. Lower-middle-class neighborhoods such as John's, whether in a small city in western Massachusetts or in Wisconsin or Missouri, would have these sorts of evidence of outdoor sports. But Jim's house only signaled the presence of young children—toys, plastic buckets and shovels, a tricycle, a wet bathing suit more flung than draped over the porch railing. Jim, his wife, and his two daughters arrived in a whirl of energy, mostly supplied by a three- and a six-year-old. Though it was nearing dusk and the yard gave ample evidence of its having been an active day for the girls, they showed no sign of wear. The same could not be said for Jim and his wife, who, among all her other responsibilities, was completing coursework for her master's degree in nursing.

We set up for the interview in the living room, but it quickly became clear that the two girls would be asking the questions, not I. Attempts to distract them with treats from the kitchen or the promise of stories were to no avail. They were more interested in what Jim and I were up to—and the older girl, especially, was eager to tell me about what she had seen in the woods when she had gone out with her dad. It was clear that Jim was going to have no trouble finding willing companions, though I was compelled to observe that his days of still hunting were clearly numbered. He shook his head as he smiled, acknowledging with a mixture of pleasure and incredulity the energy and enthusiasm of his daughters.

Jim grew up in the shadow of the Berkshires before the hill towns became an attractive spot for upscale urbanites to rusticate. He remembers accompanying his father and an older brother on some of their outings but did not begin hunting, that is, carrying a gun, until he was fifteen or so and had passed the state hunter-safety course. As he remembers it, "there really wasn't a lot to do as kid in the rural area where I grew up, so we all spent a lot of time in the woods. Some [kids he grew up with] didn't like it that much, but I really enjoyed the woods." He still does. Before his girls arrived, he spent virtually all his free time out-of-doors, either fishing, hiking, or hunting. For Jim and for several other of the contractors I interviewed, whether they were successful and had become affluent or were just bumping along, the fact that work slacked off a bit in the fall was not exactly an unwelcome feature of their trade. The lull gave Jim precious time to get out, time that he might not otherwise have had if he had a factory or office job. In recent years, with two little kids and a wife in school, he's had to hustle to avoid the lull, and as a result, his hunting has been more rushed. More important, the need to work without interruption had cut drastically into the time he has to scout before the season. This means that even when he gets out, he sees far less game than he has been accustomed to finding.

Jim was philosophical about this, understanding that it was time to shift his priorities from the field to the household, hence his eager anticipation of the day when his girls could be folded into his love of fishing and hunting. The fact that his children were girls was not an issue in the least. "Look," he said, "I'm not a feminist or anything, but I sure want to share what I love with my kids, and I hope they come to love the out-of-doors like I do." I couldn't resist prodding this "I'm not a feminist" statement a bit. Since his wife was far better educated than he was and he was actively encouraging his daughters to become outdoorswomen, his disclaimer sounded a bit hollow. Jim explained that what he meant was that he disagreed with what he saw as an antifamily stance of feminists. Even more troubling to him was his sense that feminists were allied with gun-control forces as well as antihunting and animal rights causes.

Jim was not the only hunter I interviewed who worried about the future of hunting. Most hunters, including Jim, worried about the disappearance of land suitable for hunting. Jim was one of the few who, with Peter Hanks, saw the ideological opposition to hunting as a real threat. This surprised me, for I had expected to encounter this sense of threat more in the eastern part of the state, where antihunting and animal rights sentiments are more audible. To be sure, Jim lived in one of the most liberal cities in the state, and in the county that is home to the elite liberal arts colleges, Smith, Mt. Holyoke, Hampshire, and Amherst, and the flagship campus of the state university system. The political complexion of this area is heavily influenced by the cosmopolitan values of the faculty and students who are drawn to the colleges and university. Though forests and farmland still dominate the landscape of the area, typical rural and small-town values and tastes are distinctly marginal where Jim lives. And variants of feminism are also widely propounded, such that Jim's association of feminism with antihunting and animal rights is far from paranoia.

As we talked, Jim referred to a common bumper sticker seen in the area, "Meat is Murder," as if to document the appropriateness of his concern.

He is also one of the few men or women I interviewed who had direct experience with people hostile to hunting. He recounted one incident just a few years earlier. He was bow hunting in an orchard in the town in which he had grown up, little more than a half hour's drive from where he now lives. It was an area he had hunted ever since he had begun going out with his dad and older brother. In recent years, as is the case all over New England, the farms and the wood lots are getting sold off and new houses are appearing, occupied by affluent professionals or retirees either year-round or for the summer. A former secretary of state resides in Jim's hometown, as well as numerous writers, artists, and musicians. Some of these transplants meld in and even take up hunting. Most do not. The woman Jim encountered in the orchard was not sympathetic: "She must have seen my truck parked by the old tote road leading to the orchard. I know the owner of the land, and our family's had permission to hunt there for years so I knew she had no business there. But she brought her two dogs up there and marched them around to piss on every tree. She knew what she was doing because when I came down out of my tree stand, she looked at me as if to say, 'that takes care of your hunt.'"

In another area on a different occasion, he had encountered antihunters walking through a prime hunting cover loudly banging pots and pans to frighten away game. He was certain that such clashes would be increasing, despite the recently passed law making harassment of hunters a crime. It pained him all the more to see this going on in the place where he grew up, a place that is rich with memories of neighbors helping him drag a deer out of the woods as night fell, of folks knowing who had and had not gotten "their" deer and giving the unsuccessful hunters tips on where they might spot one before the season ended. It was a place that took hunting for granted, and even those who did not hunt participated in the culture that held hunting to be part of the way of life. Not so any longer. But Jim was prepared to defend what he regards as a fundamental right to hunt.

He was adamant about this, but not wild-eyed. I probed for any hint of militia sympathy or notions of armed resistance, but there were none. Though things were definitely not headed in a direction he liked, his response was to say hunters should get involved in community affairs to make their point of view known. As he saw it, the public only hears about hunters when something bad happens. And hunters get blamed unfairly for lots of vandalism in the woods, most of which Jim thinks is done by kids raising hell. If hunters were more involved, the reputation of hunters would not be so negative. He had faith in the system's capacity to protect the rights of minorities, including hunters'. But he conceded that in a state like Massachusetts, where fewer and fewer people hunt or are tied to the land, hunting and fishing could be lost to the combination of value shifts and loss of land. If worse came to worse, Jim said, he would very seriously consider moving his family to some place more congenial to hunting. He wanted his daughters to know the wonder and excitement of hunting. He also wanted them to share something else, something he had trouble naming: "If I lost my job or if worse came to worst, there is something left that I could survive on. I always had that kind of sense, I don't know whether it's a fantasy type of thing or not, but with the knowledge that I have about hunting and fishing, there's a sense that I could survive. That's kind of a nice edge." The Osgoods said something very close to this, including the acknowledgment of the element of fantasy. Jim wanted his daughters to have a sense of self-sufficiency, a sense of mastery or control that hunting conveys. Had he known Karen DeFazio, he would like to think his daughters would grow up to be like her.

Karen is an energetic mother of five who looks to be in her early thirties, but that would mean she had her first child when she was ten. Good genes and a very active life have combined to keep her youthful. Karen is a special education teacher with advanced degrees in both education and psychology. Her husband is employed in the high-technology industry that dominates the economies of the towns and cities between Worcester and Boston. They live on a wooded lot little more than a stone's throw away from the busy freeway they both take to their respective jobs. Until last year, the back of their lot abutted a large wooded area in which both she and her husband had hunted. He had taken a deer with a bow several years before, and the mount of its head with a nice, eight-point rack graced the wall of the den. But as we spoke, the sound of chain saws was a constant backdrop. The wood lot was on its way to becoming yet another subdivision. When they learned of what was going to happen, Karen and her husband put the house up for sale, thinking that they might be able to find something "further out, with some acreage," something like an old farm with enough land to provide them with hunting, come what may. But at the time, the economy was not particularly robust, so they took the house off the market and are reviewing their options. "Maybe when the kids are a bit older," Karen said.

The dream of owning a place of one's own where family and friends could freely hunt without having to worry about development pressures or unsympathetic landowners was a common theme among the hunters with whom I spoke. As Karen reflected on the ways hunting opportunities are getting more constricted, she said "I would have loved to have been born 150 years ago. I think I was born too late . . . I would have been a great pioneer woman." As it was for the Osgoods, Jim Ramada, and a number of other hunters, Karen's sense of connection to the past is linked to a sense of herself as capable, resourceful, and autonomous. At least for those who hunt, hunting seems particularly rich in its capacity to confirm skills and personal qualities we associate with pioneers, subsistence farmers, and the flinty, taciturn characters that Norman Rockwell's or Grant Woods's paintings help us imagine. Karen was proud, if a bit overconfident, about her ability to keep things together even if the rest of the world went to hell. "I know without a doubt that I could feed my family, all five children. It would be no problem at all. That gives me a real sense of satisfaction, knowing that I know how to take care of things."

Karen began acquiring this knowledge early. Her father gave her a small-bore shotgun when she was ten, and she has hunted ever since, though she graduated to a 20- and then a 12-gauge as she grew up. She accompanied her father and other members of the family on their annual deer hunts, but those outings rarely involved more than two or three days a year (when she was young, the deer season in Massachusetts lasted only ten days). As a result, most of her hunting was for birds, primarily pheasants, and this remains the case. They would head out from Boston early on Saturday mornings in October and November to hunt pheasants on the farmland that was still abundant right around where she and her husband now live. In fact, there is one large parcel still under cultivation just down the road from her present house where she and her father used to hunt. It is still hunted, but the pressure is so great that Karen no longer cares to hunt there, even though it is handy.

> A lot of people come in from Boston to hunt in those corn fields. I no longer hunt there because there are hunters that don't know how to hunt. I used to bring my children down there to start them off with bird hunting because it was so convenient and because the state stocked birds [pheasants] so there was a good chance of the kids having an opportunity to shoot. The day I had to tell one of my kids to duck because some goof was shooting toward us was the last time I hunted there.

For many years, Karen and her family have been hunting on privately owned land in western Massachusetts, a little more than an hour's drive from her suburban home. The land is posted, but the family has had the owner's permission to hunt for as long as Karen can recall. It is land with which she is intimately familiar: "I can come over a crest and know which trees are there. I know where everything is and that's part of the joy of being out there." Because the land is posted, it is also reassuring to know that you know who is hunting there. The same group has more or less hunted together for twenty or thirty years—they know the land and one another's styles of hunting (some prefer stalking, others prefer sitting) so the chances for accidents are very small. And since all are friends, there is no competition of the sort that compels some to take risky shots or to argue over who shot the deer or bird.

Hunting alone in a public hunting area two years prior to our conversation, Karen had just such an encounter. She and another hunter unknown to her, who she described as an "older gentleman," her voice filled with irony as she uttered "gentleman," both shot at the same deer. The "gentleman"'s shot struck the deer in a hind leg; Karen's shot was to the lungs, a killing shot. The "gentleman" rushed to the deer and claimed it for himself. "I told him. I said that it was my shot that killed the deer and I showed him where the deer was hit, and I went back and got my empty shell and showed him the angle and retraced the whole thing. It was clearly my shot and my deer. He just shook his head and took out his knife and began dressing the deer. I knew I had to get out of there." Experiences such as this have not diminished her love of hunting, though she did quit deer hunting for a couple of years after a different kind of problem befell her. We were talking about the code that dictates eating what you kill, and she related this anecdote: "I really am adamant about not shooting something that I won't eat. It's not the kill. That's no fun at all. It doesn't bring me any joy. In fact, there was a time when I was with some hunters and they shot a doe by accident, and we gutted it, put it in the woods, and we knew we couldn't pull it out because no one had a doe permit. We left it there and I stopped hunting for two years. It was so traumatic . . . I was just devastated."

Experiences such as these are, so far as I can determine, uncommon, but they are frequent enough to lead some hunters to call for better training of hunters and tighter regulations. Even though these two experiences left a lasting impression on Karen, she was not inclined to seek tighter controls.

In fact, she was worried about all the regulations now on the books, not only about hunting but about seat belts, helmets (Karen rides a motorcycle, though with less and less frequency in recent years), "everything—I think it's bad, too heavy. Too many eyes looking at me." Part of the attraction of hunting occurred to her as she registered this concern. "When I am hunting I'm truly by myself. Nobody is there. Nobody there to look at me. it's all my responsibility—I hold all the strings. It is totally up to me whether to shoot or not. I am in charge."

It would be easy to read these words as a macho script or a power trip. But that would be a profound misreading of what Karen was getting at. She was worried we were becoming so hemmed in by regulations that no one feels responsible for his or her own fate. She was not asserting her control over other humans or even control or dominance over the animal kingdom. When she said, "I am in charge," she meant that she was in charge of herself, for the moment beholden to no one. Hunting thus puts her directly in touch with herself.

Karen and her husband hope to pass hunting on to their children so that they, too, can become sturdy and self-reliant. Their oldest two, a boy eighteen and a girl fourteen, have tried their hand at hunting and not found it to their liking. Ironically, the difficulty may have arisen from Karen's strong conviction that you have to eat what you kill. Neither likes to eat the game Karen and her husband bring home, so given the announced morality, they should not hunt. The younger three, by contrast, like game, at least so far, and while still a bit too young to be taken out, all three, Karen said, look forward to the time when the little .410 their grandfather gave to their mother is placed in their hands.

Robert Swipe also has a gun waiting for a youngster to be old enough to be taken hunting. Tall and powerfully built, Robert Swipe has close-cropped, thin, greying hair, the only feature suggesting his age (sixty). Robert has three brothers and three sisters from whom he was separated for most of his childhood and teens. His mother had been abandoned by her husband, and she simply could not manage. Robert and several brothers and sisters were raised in state-run homes for the indigent. He was obliged to work on weekends and during school vacations, with most of his earnings taken to defray the costs of his room and board. He was a city boy, but, perhaps because of his build, he was assigned to work on farms during the summers and got a taste of country life, albeit as an "indentured laborer." He remembers being taken hunting, beginning when he was fifteen or sixteen, by the husband of one of his older sisters. He was, he admitted, much less interested in the hunting than he was in the gun he was allowed to carry, a .22 caliber rifle. It was a poor introduction to guns and hunting, as he recalls. Safety "was something I learned on my own, I guess, because I don't remember him ever telling me anything about it. As a matter of fact, I got shot. One of the guys we were with must have shot at something and missed and got me. That's how I learned to think about where your bullets go." Fortunately, he wasn't badly hurt.

The incident did not dim his ardor for guns. Of course, while he was living in the institution he could not own a gun, even if he could have scraped together the money to buy one; and when he was reunited with his mother when he turned eighteen, she forbade him from owning a gun. He continued to go out "hunting" with his brother-in-law, but he spent most of his time shooting at tin cans. He can't remember ever even shooting at an animal in those early years. One of his first jobs when he was back with his mother was at a slaughterhouse, and there he shot his first living animals (a .22 to the head was the method practiced). He remembers being troubled about this for the first few days, but then killing and butchering ceased bothering him. Still, this experience did not whet his appetite for hunting. When he was twenty-one, and no longer needed his mother's permission, he bought his first gun, and he and friends would go on their free time to local shooting ranges or just out into the country to shoot at targets but not to hunt. He got interested in hunting when he was well into his twenties.

Robert had been working two and sometimes three jobs in order to save enough money to secure the release of his younger siblings who were still wards of the state. One of them was assigned work with a woods crew, many of whom hunted. The youngest was intrigued, and when he was reunited with his family, he convinced Robert to go along with him. With no one from whom to learn, the two blundered along. "I didn't know anything about hunting. Nobody ever told me. I read some books, and I listened to guys talk about how they did this or tracked that. And in the woods I started putting things together a little bit. And so I started going with my brother and teaching him, learning myself actually, and I began to get serious."

He laughed at himself as he shared the memories of the first deer they shot.

My uncle had a farm in New Hampshire with eighty acres that we started hunting on. The farm had plenty of deer. I was so excited when I shot my first deer that I walked off the tree I was in. I was up in a tree a ways away from my brother, and I saw this deer coming right between us. I waited until he was past my brother, who was on the ground and didn't see the deer, and then I shot. The deer bolted and ran toward the swamp. I didn't know if I'd hit him, and I hadn't learned that you should never charge after a wounded deer. But I forgot I was in a tree, and I took one step and whamo, I fell. Broke the stock on the gun, but luckily it didn't go off and I didn't get more than shaken up. We followed the deer into the swamp and found him near dead. I finished him off. I still can't believe I walked right off the friggin' tree.

Robert is, in fact, self-taught at almost everything. He was, he said, no good in school. He just couldn't make himself study things that did not seem practical or immediately applicable to the overwhelming need to make a living. He left school before graduating (though he subsequently got his high school equivalency certificate) and began working. As already noted, among his early jobs was a stint in a slaughterhouse. Handy with his hands, he also began to establish a reputation as a pretty good shade-tree mechanic and welder, skills he acquired on his own. He also taught himself how to repair small appliances. His incredible work ethic and his ability to quickly figure out how things work ultimately landed him an almost unimaginably good job with a company that did specialty machine work for defense contractors. He eventually became manager of the works, which not only paid him well but also meant the firm provided him with a car, all expenses paid, and a number of other perquisites. During the Reagan years, the defense industry was booming, and Robert's fortunes rose. But, when Robert was age 55, the cold war ended, defense contracts shrunk, and Robert's bubble burst. A senior executive from corporate headquarters gave him the news: his position was terminated, and since his service with the corporation was just shy of the twenty years the company required to trigger a golden parachute, he was let go, on the spot, with no pension, health coverage, or life insurance, and no car. Two months' worth of severance pay, a handshake, and barely time to clean out his desk was his reward for the hard work he had given the firm.

He had saved—his house is modest in a neighborhood of modest, middle-income families in Springfield, the largest city in western Massachusetts—but just months before that fateful day, he had tapped almost his entire savings in order to set his daughter up in a beauty salon. The daughter apparently did not inherit Robert's drive and acumen—by the time I interviewed Robert, five years after he had lost his job, the daughter's hair-dressing business was bankrupt, he was working two part-time jobs in local liquor stores for minimum wage, and paying off utility bills left from his daughter's failed business. He was virtually back where he started—repairing cars, doing a bit of welding, and working on appliances. Ironically, he hunts less now than when he was working steadily because he has no vacation time, his work schedules are such that he has no extended block of time in the day, and he has to hustle to keep the small repair jobs coming in. He dreams of saving up enough money to start a business of his own selling guns—his lifelong hobby.

In the two years prior to our interview, Robert had gotten only one deer, but the skills he had acquired long ago in the slaughterhouse pay off: friends and acquaintances bring their deer to him for butchering. As is customary, the person who does the butchering is allowed to keep a portion of the meat, so he has been able to relieve a little pressure on what has to be an extraordinarily tight family budget. But Robert does not think of himself as a "meat hunter." Instead, he says he hunts for the pleasure of getting away from the concerns of life—when times were good, hunting took him away from the pressures of a demanding job. Now hunting takes him away from the tensions and anxieties of barely making ends meet.

But Robert spoke of hunting almost in the past tense—though vigorous, he tacitly knew his hunting days were nearly over. His only son had accompanied Robert when he was a teenager, but his heart wasn't in it. He went along, Robert implied, because he knew how badly Robert wanted to be a good father, the kind of father Robert never had. "He was going with me to go with dad, and I wanted to be with my son hunting. I bought him a gun and he hunted with me every year for a while, but I began to see he didn't like it. I'd come back to the Jeep and find him sound asleep. And even though he's a very good shot—I made sure he really learned about guns—he'd never shoot, never fired a shot. So after awhile I just said to him, 'you don't have to come with me if you don't want to.' So he hasn't gone with me for the past eight years." When his son moved into his own place, he didn't take the guns Robert had bought for him.

The group with which Robert hunts is moving on too. For years, between twelve and fourteen guys would hunt, not all at the same time but over the course of the season, in pairs or groups of three or four; all would share one or more hunts together and get together after the season to keep the memories fresh. Now, the group is down to three or four. Some have died, others have retired and moved away, and others simply put their guns away. With less time and a lot less money, Robert wasn't even sure if he was going to buy a hunting license for the coming fall season. The social supports that sustain most hunters are slipping away from him. Even though he was clearly unhappy with this prospect, he was not morose. He has a grandson and is looking forward to introducing him to hunting when he gets old enough.

There's a slight problem, though. Robert's daughter hates guns and hunting and expresses sympathies for animal rights. For a while she refused to let her child visit his grandparents because of the guns her father owns, even though Robert's many guns are all securely locked up and kept out of sight. Somehow, her objections have softened, perhaps because Robert and his wife are only too happy to care for the youngster, an attachment made evident by the myriad toys and the prominent place his high chair occupies in the kitchen. Robert told me that he has bought a shotgun for the lad, against his daughter's wishes. He relishes the prospect of going out with the boy. "I can't wait to take my grandson, even though my daughter says he isn't going out. Hunting wasn't passed on to me the right way, but I think I can introduce my grandson to hunting the right way, let him see what the woods and animals are like. He should learn to shoot, learn all about guns and hunting and decide for himself if he likes it. Like my son. And if he doesn't like it, that's okay. At least he'll know something and, who knows, maybe later in life he'll take it up on his own, but he won't have to learn the hard way, like me."

Maybe the grandson will take to hunting and Robert will get a chance to pass on something that has been very important to him. Guns and hunting have come to embody Robert's sense of accomplishment and his indefatigable pursuit of the solidity he was denied as a youngster. The rug was pulled out from under his self-made-man saga, but the skills, knowledge, and adeptness that he has acquired with guns and hunting can't be taken away. They are his true possessions, the only things that he is likely to be able to bequeath.

Unlike Robert's daughter, Elaine Stebbins grew up in a family that was antihunting ("not actively antihunting; they just don't like the idea of hunting"). But there was a bit of woods beyond her backyard in the Boston suburbs, and she recalls frequently playing there with her friends. Those childhood experiences probably were responsible for sparking her early interest in science, especially biology. She majored in biology at the

University of Massachusetts, and it was in a biology class that she met the man who was to become her husband. He was majoring in forestry and biology was a requirement. Before long they were courting, and during the courtship, he took her along on a couple of bird hunts. She was intrigued. He got her a shotgun for an engagement present. She smiled as she told me this, as if to say that hers was a different kind of "shotgun marriage."

They were married shortly after graduation and moved to the Berkshires, near where her husband had grown up. Jobs in forestry were scarce, but his local connections made it possible for him to start his own small general contracting business. Elaine started out working for an environmental consulting firm, where she worked for six years before leaving to start a business of her own, operated out of their home. When she told me this, I assumed that like so many people these days, she was doing consulting work on her own, connected to clients via the electronic office. After our interview was concluded and as I was packing up my tape recorder and notebook, Elaine asked, "Don't you want to know what I do?" The question took me aback. How could I have let this slide? "Sure," I said, hardly concealing my embarrassment. She invited me to join her in the basement of her modest frame house. The basement was crowded with the skins of birds and animals in various stages on their way to being "stuffed." Elaine explained how, in her spare time, she had been taking taxidermy courses by mail, and then, when she became confident that she could do it well and make a living at it, she enrolled in a two-month course in the Midwest and returned to start what quickly became a brisk taxidermy business.

What Elaine had discovered about herself, beginning in the woods behind her childhood home, was that her interest in biology was really more nearly an interest in animals and natural history. In this sense, she shared a passion for the detail and uniqueness of each species with the bird carver, Bill Crafter. Her hunting allowed her to see animals in their natural settings—to see grouse in flight, to closely observe the posture and bearing of a buck or a doe, so as better to represent them in her mounts. As she described her work, it struck me that she had managed to contrive a kind of seamlessness between her vocation and avocations that most of us can scarcely even dream of.

As for many of the people I interviewed, the idea of living off the land was enormously compelling for Elaine. She and her husband maintain an extensive vegetable garden in the back yard and also keep several chickens which keep them supplied with fresh eggs. "We eat meat almost every day," she told me, "but I can count on the fingers of one hand the number of times we buy meat at the store." To be sure, even though her income dropped sharply when she left her job with the environmental consulting firm, Elaine and her husband didn't have to hunt in order to have meat on the table. But like the Osgoods and Karen DeFazio, Elaine conveyed a sense of satisfaction with the degree of self-containment she and her husband had achieved in the scant seven years since they had graduated from the university. Indeed, Elaine seemed genuinely torn between her desire to see animals up close, to observe them minutely, and her desire to eat them. In the end, she confessed that hunting combined both these desires of hers, giving her "the thrill of getting your own food."

The Stebbinses' house is the last house on the street, surrounded on three sides by state forest and a state wildlife management area. Now that she works at home, Elaine said with evident pleasure that she can take a break from the intense demands of taxidermy, pull on her boots and hunting vest, and walk out the door into prime cover. She and her husband hunt both small game (upland birds, waterfowl, and rabbits) and big game (turkey, deer, and bear; neither of them has yet to get a bear—"I've seen several but just didn't have a good kill shot," Elaine told me). They use bow, shotgun, and black powder, which means that they can hunt from mid-October to mid-December for a mix of birds and big game; through February for rabbits; and then for two weeks in May for turkey. Unlike a lot of other hunters I interviewed, Elaine does not hunt out-of-state, and she showed little inclination to even think about doing so, even though her husband has begun to hunt deer in New York State.

Elaine's love of hunting is something of a mystery to her parents and siblings, and it has led to some aggressive questioning. But aside from that, she has personally encountered no flak regarding hunting. The men in the rod and gun club Elaine and her husband belong to are quite welcoming, and she has never had anyone be rude or insulting to her in the field. She has heard of some antihunting protests in the area but has never witnessed any herself. It was not surprising, then, to learn that Elaine was not particularly fretful about the challenge of antihunting sentiment. She was, however, worried about the impact continued population growth and the dispersion of people outward into the countryside from the city was having on hunting. Though she had lived in the Berkshires for less than a decade, and those few years were not banner years for home building in western Massachusetts, she has nevertheless already seen the loss of some fine hunting areas to development pressure. Unless something is done, she was pretty sure that hunting would be less and less satisfying because the variety of covers would decline and hunting pressure on the few areas that remained would intensify to the point of ruining the experience.

Elaine is not a joiner or an activist, so her response to this gloomy prospect was entirely personal—comfort in knowing that the area immediately surrounding her house would be spared because it was state-owned land dedicated to conservation-use only; cultivating land owners in the area who will grant her permission to hunt even if they close their land to the general public; and resolving to have no children, thus not contributing to the population problem. And there was a last resort that has a familiar ring: "If it gets real bad, I think we would have to think about moving, maybe to Montana. My husband and I do talk about going there for a vacation, to hunt and fish, but also just to 'check it out.'"

Jack Wysocki, the youngest of the males I interviewed (Sioban Osgood was a couple of years younger than Jack), has checked other places out, including Montana. He lives in a town whose rich farmland is rapidly disappearing beneath malls and subdivisions. Jack has yet to find a way to fully accommodate to this change. On the one hand, the steady development of residential and commercial property has meant plenty of work, especially for someone like Jack, who is not particularly drawn to books and studying. He lives at home and has worked at a variety of jobs all related in one way or another to the transformation his hometown is undergoing. At the same time that he is grateful for the work and applauds the entrepreneurial spirit that has been kindled, he laments the steady loss of hunting opportunities and, more particularly, the loss of many of his most treasured coverts. Like Sioban, Jack began hunting early, first accompanying his father and uncles, and later, when it was legal for him to go unaccompanied, he remembers coming home from school, changing into his hunting garb, and going "just across the road to hunt for rabbits and pheasants." His hunting is now somewhat more elaborate. He and his father have been to the Rockies in pursuit of elk, they've hunted moose in Canada, and they have plans for more such trips. The den through which one enters the raised ranch–style house is filled with mounted creatures—a flying turkey, a bear, and red fox, and numerous heads of deer, elk, and a moose, at least fourteen in all. Unlike the Osgoods' display, which struck me as akin to a natural history exhibit, the mounts in Jack's and his father's den were there not so much to feature the animals as to proclaim Jack's and his dad's exploits.

Jack still likes hunting locally the best because the coverts that remain are filled with warm memories, but the West calls to him. As he described his last trip west for elk, he noted that he had always liked Westerns and that figures like John Wayne loomed large in his imagination. Hunting, he said, makes him think of the cowboys, the scouts, and the Indian fighters who fired his imagination as a child—and still do. Though not the studious type, he has carefully studied and mastered the fundamentals of wilderness survival. He reeled off a host of edible plants one could rely upon if need be, and his volunteer work with fire and rescue crews has given him a wealth of practical knowledge about survival as well as self-confidence to endure physical challenge. I had the distinct impression that Jack would like nothing better than a full-blown natural disaster against which to test himself.

Well, I think he'd like one thing better: the regular opportunity to hunt in wide-open "big country," where adventure and challenge add to the richness of fantasy attached to such trips. In unfamiliar country, he can test his skills, his endurance, and his savvy—he can affirm his manhood in precisely the archetypal ways in which he thinks. Hunting, for Jack, is associated with the heroic chapters in American history when the West was "won." He would have liked to have been there, for some of the same reasons that the far more mature and thoughtful Karen DeFazio would have. Though he did not say so explicitly, he clearly likes life on the "edge," a life filled with opportunities for decisive action and heroism. Work with the rescue squad, including underwater search and rescue, helped meet this need, as did his volunteer fire crew activities. Hunting provides Jack with another way to be a man of action, and it gives him a way of being in, but not fully engaged with, the society around him, a society that rewards contemplation more highly than action. Historians following Frederick Jackson Turner may agree that the frontier closed more than a century ago. But the wide-open spaces of the West seem a sufficiently large canvas on which Jack now imaginatively projects himself. Like generations of Easterners before him who found the settled East too confining, Jack may before long break out; the loss of a prized covert will be just the impetus he needs to move to the West. For Jack, the search for new hunting grounds will be as much for an imagined past as for a future.

III

A LIVING PAST

Bound by a thousand tiny threads like Gulliver in the land of Lilliput, hunters are linked to the past. For some, these threads are woven into cables that bind them quite closely to an agrarian way of life in which hunting is integral. For others, only a thin thread or two connects them to the past. Much like the people who preserve railroad culture in their elaborate scale models of legendary trains or those who go to great lengths to reenact Revolutionary War or Civil War battles, hunters, too, are engaged in reenactment. To be sure, the analogy is a loose one since there is no one script hunters follow. We know, more or less, both the broad contours and many of the details, down to the letters and diaries of specific individuals, of historic battles. By contrast, the threads linking hunters to the past are different for each person, not least because each has his or her own version of the past to which he or she is drawn. This does not, however, diminish the hold the past has on hunters' imagination. If anything, it makes the past more compelling.

Ties to the Land

Several of the hunters I interviewed were raised on farms and have close relatives who are still actively farming. They return each fall to the farms where they grew up, if possible several times, to hunt. In one case, the pull is more the farm and kin than the hunt; for others, the magnet is hunting with kin. Keith Jones exemplifies the latter. Keith was raised on a farm in upstate New York. After graduating from high school, he joined the Navy and was based in Fall River, where he met a young woman whom he married. When his hitch was up, they stayed in Fall River, where he used the training he'd gotten in the Navy to set up a small electrical contracting business. The couple had three children, but soon after the third child was born, Keith's wife ran off with another man. Keith has raised the children by himself as well as kept his business afloat. He is an energetic man in his early forties and he loves to hunt. But being a single parent and independent contractor doesn't leave much time for hunting. As a result, he rarely hunts in Massachusetts, though he buys a license each year "just in case the chance comes up," and eagerly looks forward to the day when he will have more time to hunt nearby. For now, though, he has to settle for one extended outing a year. He packs the kids in his van each November and heads back home for a week or ten days of deer hunting with his brothers who have taken over the day-to-day operations of the farm. Keith's father joins "the boys" for a morning or an occasional day, but Keith noted, with a touch of regret, that his father's interest in hunting has waned in recent years.*

Keith's interest has certainly not flagged, and he has fully conveyed his enthusiasm to his two sons. I arrived at Keith's before he got home from work, and when I rang the doorbell, his daughter invited me in. She and her two brothers were getting the evening meal together, and as they went through their respective chores, we chatted. The boys talked animatedly about how they loved the trip back to the farm and how eagerly they looked forward to the day when they would be allowed to accompany their father and uncles on the hunt. Keith's daughter, the oldest of the three, had to remind her brothers that they had work to do, as it became clear that the boys would rather talk to me about hunting than peel potatoes and trim green beans. When one of the boys left the kitchen to fetch the BB gun

---

*This is a pattern often remarked upon in the literature on hunting. In some small measure, the decline in participation with age may reflect a loss of vigor and stamina or a disinclination to put up with the inevitable discomforts that accompany hunting, especially deer hunting. But clearly more is involved. For now we might think of it as a loss of "appetite," not for the meat but for the kill. I explore this more fully in chapter 4.

he'd gotten for his twelfth birthday, she good-naturedly called out to him to ask if he intended to go into the back yard and shoot some song birds for dinner. She made it clear that she enjoyed the annual family reunions, but she screwed her face up in disgust when I asked her if she had any interest in hunting. She eats venison and other game her father brings home, but not with relish. Later, when Keith and I were talking, she registered a playful "yuk" when he was relating the account of how he'd shot his biggest deer. With equal good humor, he chided his daughter about becoming a Bambi lover.

Keith misses the closeness to the land that he experienced as a boy. But unlike some of the urban hunters who dreamed of owning their own small farm, Keith knew full well that he was not going to return to farming. "No future in it," he said matter-of-factly. But there is a past in it, and Keith means to keep that alive. His dining room is decorated with farm memorabilia, miniature milk cans, carved wooden cows, and similar bric-a-brac. Living in one of the oldest mill towns in the country, with little room and even less time for so much as a garden, Keith has made the most vivid and rich connection to a way of life he cherishes through hunting. Returning to the family farm, reuniting with family, and hunting on land with which he's intimately familiar makes the rest of a very hectic year bearable. "What if the farm gets sold," I asked him. He paused, and his children stopped their muted banter and turned to listen. "Well, it's going to happen one of these days," he said. "It's too hard to make a living at it. I'll miss it, but we'll still get together to hunt in the fall. Farming is not in my blood, the way it is for my dad, and it's not the same for my brothers either. But hunting is in my blood." The boys were relieved. The spunky young girl shook her head in mock disbelief, as if, wise beyond her years, to say "that's men for ya."

Andy Felter also grew up on a farm, in the Berkshires of western Massachusetts, and like Keith, he returns to his each fall. For this skilled technician in a computer firm, the trips back to the farm are his only occasions for hunting. Though there are some limited hunting opportunities nearer his home in the Lowell-Lawrence area, and the woods of New Hampshire are a short drive from his home, he doesn't like hunting anywhere but on or near the farm. As this might suggest, Andy isn't a very avid hunter. Unlike most of the other men and women with whom I talked, he wasn't itching to get into the woods. He was itching to be home. It struck me, as I was making the two-hour drive back to Amherst after our interview, that hunting was an acceptable way of concealing homesickness from himself and from others. He had little interest in guns, confessed that he did not like to be in the woods when it was dark or foggy or cold, and had trouble recalling a particularly vivid or exciting hunting experience. In fact, the only thing that stands out in his memory is the time, just after he got his first .22 rifle, that he shot his mother's cat while experimenting with his new gun. He has yet to get a deer, and I suspect he never will. Andy hunts for tradition, not for game.

Not surprisingly, Andy and his wife would like to move back to the Berkshires, and were he to find a job there, they would do so in a heartbeat. Andy's home suggested his reluctance to set down roots in suburbia, a reluctance his wife made clear she shared. Though he earned a decent salary, he had not invested much money or time in the house. It was not run down, but it lacked the signs of care that other places on the street boasted—carefully tended flower beds, a deck, or groomed hedges. Andy's was a house, not a home. He ruefully noted that there just weren't jobs back in the Berkshires for people with his training and expertise and, like Keith, he had no illusions about making ends meet as a farmer. Indeed, his father had long since given up active farming. He just mows the hay fields, keeps a few chickens, maintains a vegetable garden, and harvests a couple of cords of firewood each year. So Andy and his wife are stuck in a place they'd rather not be, a two-and-a-half-hour drive from where they wish they could live. The fall ritual enables Andy to walk the land he loves and keeps alive the prospect of a time when he might return permanently. "Do you think you would hunt more often then," I asked. He paused before saying, "I'm not sure," as if to say that were he back on the farm, he wouldn't need the hunt to connect him to the land.

Paul Julien is an independent general contractor who was enduring an uncomfortably long period of unemployment when I interviewed him in the middle of what should have been his busiest season. Paul's extended family is based in northern Maine, where they grow potatoes. As a youngster, he spent summers on the farm. Now, he mostly goes back to visit, though when the need arises, he goes back to help out. Sometimes these visits are timed to include hunting, but unlike Keith and Andy, Paul doesn't need to go home to hunt. Paul hunts to stay connected less with his own past than to a time when settlers and natives commingled, if not always peacefully, and learned from one another. He can walk out his back door and begin hunting as soon as he is a safe distance from the road. Although

Paul talked about how hunting kept alive his links to the land, he is much more interested in the ways hunting moves him to think about the different ways Native Americans and colonists related to the land. Though he prefers to use a shotgun ("because it gives a better [i.e., quicker, cleaner, surer] kill"), he spends a lot of time reflecting on Native American culture and marveling at how the natives managed to live off the land. He admires their ability and the world they made for themselves. He has become friends with a Native American in the area who hosts "sweats" and other ceremonial gatherings in which participants experience a radical intensification of emotional bonds to nature that parallels his exploration of simpler styles of life.

Paul's economic woes might suggest that a simpler life has been thrust upon him, but that would be a mistake. Paul was clearly not enjoying being out of work (though I dare say he might have been less concerned had it been November rather than July), but there was plenty to do around the house and in the extensive garden he and his wife maintain. Paul combines the refined skills of an experienced builder with an eye for the found object. To make something of salvage materials otherwise headed for the landfill and cleverly to incorporate them into something for which they weren't intended gives him great satisfaction. It's not so much saving money, though that is not irrelevant; Paul is pleased because he feels that he is closer in spirit to those who, long before him, lived off the land. From his perspective, it is hardly a stretch to enlarge this formulation to "living off the landfill."

Paul hunts with this same orientation: hunting is part of an elaborate complex of activities that, together, describe an ideal toward which he aspires: a life that as near as humanly possible allows him to balance what he takes from the land with what he returns to it. He is no fool however: he knows that the way of life he admires is gone, and he cannot will it back. Still, he chooses to hunt with no gadgets and no frills. He would like to get adept with a bow, but he isn't nearly good enough, by his own lights; he refuses to risk wounding an animal and losing it. Unless he is hunting in unfamiliar country or in the really big and largely trackless stretches in northern Maine, he goes out without a compass, confident that if he were to "get turned around," he would know how to forage and to keep body and soul together long enough to make it back. "Getting lost," he told me, "is really just a scheduling problem. If I don't have to be anywhere at a particular time, and I can feed myself and stay warm, then I'm not lost."

Paul and his wife, like many of the men and women with whom I talked, maintain a large vegetable garden and are serious about composting—"not much goes to waste around here," he noted with satisfaction. It is ironic that someone who makes his living demolishing buildings and building new ones would, in his own life, practice an ethic of minimal impact, but there it is. Hunting and gardening keep Paul close to the land; his handiwork with boulders, construction "debris," and gnarled tree limbs and other "useless" objects completes the ethical circle he has drawn for himself. When he hunts, he tries to hunt as he imagines a Native American might have hunted.

Unlike Native Americans, of course, Paul does not face starvation if he does not come home with a deer. In fact, though he is avid about hunting, he does not get a deer every year. By all available measures, he is far luckier with debris—his modest house is graced by all sorts of amenities that would be far out of reach were they not salvaged. We met and talked on a warm summer morning behind the house. Paul had excavated back into the hillside that his house would otherwise have hugged. Instead of creating a yard, he had used boulders from the hillside, as well as some hauled out of the earth when he put in the garden, to cover the scar of excavation with a stone wall. With a canopy, the area had been made into an outdoor "den." When I arrived, he was just finishing a leisurely breakfast of pancakes and coffee prepared on a wood stove that is the central object in the "den." The lazy wafting of wood smoke kept the insects at a respectful distance, and even though we were but five yards from the house and not more than forty yards from the road, it seemed as though we were back in time, perhaps in some logging camp in Maine. The only incongruous objects in this scene were my tape recorder and the bright orange extension cord that fed it electricity.

Paul is soft-spoken, a surprise at first because he is a large-framed, muscular man with a luxuriant, full beard. I expected a booming voice to match his strong handshake. Instead, he spoke with slow deliberateness, pausing to reflect so that he would choose just the right words to express how he felt. Like all but a handful of the people I interviewed, he could relate past hunts with precision, down to describing the tree from behind which a buck had emerged into the open or the streambed into which the deer he was about to shoot last fall vanished as if into thin air, "when we really could have used the venison." Each example illuminated some feature of nature or of wild animals that Paul found compelling. To be good enough, smart and knowledgeable enough, to be able to have even the chance of a good shot was something to be proud of. In just the same way, he was clearly proud of his handiwork. He was feeling the pinch of not having any money coming in, but he was in all other ways proud and confident because he was living in ways that time had tested. He did not use the currently fashionable term "sustainable" to describe either the past or the way he was living, but that certainly would be an appropriate term, at least to describe his intentions. Hunting was integral to this way of life, then and now. Being able to keep the historical thread intact gives Paul more than satisfaction; it gives deep legitimacy to who he is. It grounds him in a way that others might experience through religion. It helps answer the question, Who am I? as well as providing the measure against which he affirms his self-worth. Not coincidentally, it means that unemployment, although unpleasant, does not threaten his self-esteem.

For the most part, the people raised on farms had the strongest ties to the land, but a surprising number of hunters with no direct experience of farming expressed deep attachment to the land on which they hunt or have hunted. They return each fall to hunt in areas that hold potent sentimental meaning for them—an area they hunted with their father when they were youngsters, the place where they or their fathers or grandfathers got legendary bucks, or a covert where a hunter's first bird dog made a memorable retrieve. As is the case for hunters in general, many of the men and women I interviewed grew up in small towns where, thirty or forty years ago, hunting was commonplace and easily accessible.

Rob Collins is a good example of someone who grew up in such circumstances. Until his graduation from high school, he lived in a small town in north central Massachusetts. He wasn't exactly sure when he began hunting, but he remembers that his dad gave him a shotgun, a single-shot, when he was thirteen. That's when he began hunting on his own and with friends in the ample open spaces in and around town. After earning a bachelor's degree, Rob joined the Peace Corps, and when his two years were up, he remained abroad, mostly in Southeast Asia, working as an economic development consultant. When he returned to the States, he earned a graduate degree in planning and has worked ever since as an economic development planner in Massachusetts, helping to breathe life into local economies rendered nearly comatose by the flight of manufacturing and the collapse of computer giants like Wang and Digital. In connection with his work, he has advised a small manufacturer who is interested in resuming production of single-shot shotguns like the one his father gave him, the gun he continues to use. He is clearly pleased that economic recovery may be boosted by industries using modern production techniques to produce replicas of guns in use a century or more ago.

Rob has hunted big game in Asia with high-powered rifles and is no stranger to technical talk about guns. In the past, he enjoyed shooting, but he no longer does much target practice. Not as devoted to hunting as many of the hunters I interviewed, but certainly more interested in hunting than Andy, Rob nevertheless fondly remembers bringing home a steady supply of pheasants and other small game when he was young. Now, though, bringing game home is not important to him. Rob hunts almost exclusively in order to "stay in touch with the land," particularly the lands on which he hunted as a youngster. Though he occasionally hunts with friends elsewhere, most of his time afield is spent in familiar haunts. Getting out, inhaling the distinctive air of the New England fall, submerging into the foliage, all of these bring back memories that Rob treasures. It would seem that he would not even need to carry a gun, since bagging something is hardly what pulls at him. And yet, he notes that it "just wouldn't be the same' without a gun. It is almost as if his gun were living and in need of exercise, and Rob was taking his gun for a walk. The gun somehow concentrates and intensifies the experience, even when it is not fired. Merely being capable of shooting is enough for Rob now. Carrying his old single-shot concentrates things even further. Having a single-shot imposes a discipline and focus much like that required of an archer. It also makes the odds more nearly those that his forebears faced and implies a frugality that is reminiscent of the constraints agrarian and working-class folks, such as those among whom he grew up, face. In these very specific ways, Rob hunts to relive the past and remind himself of the social distance he's traveled: from a working-class boy in a parochial community to a cosmopolitan and accomplished professional.

Zane Zacharias has never left the town in which he grew up—a town that is a patchwork of forests and farmland—except to attend the state university. A recent graduate in his mid-twenties, he has yet to find his vocation. He likes physical activity and the out-of-doors, so until something better comes along, he works seasonally for a local landscaper and picks up a construction gig now and then. The advantage of this line of work is that the landscaping business ends just as the fall hunting season begins. His unemployment is welcome. He is unmarried and lives with his father, so he can get by on his modest income. But even if this lifestyle made getting by really hard, he would likely persist, at least for a while.

Zane grew up with hunting, and he clearly loves to hunt. Everyone in his family and nearly everyone in town, at least the male population of the town, either hunts or has hunted. It's the kind of town in which boys disappear from the junior high and high school during the two weeks of deer season. Hunting, for Zane, is a thread that is woven through virtually everything he has done and everyone he has known. Though less reflective about these matters than many of the older hunters with whom I talked, Zane finds hunting the most palpable way to pay homage to his ancestors. If he does so in few other ways, at least in this way he is honoring the traditions they lived by. Much like someone placing flowers at the gravesite of a beloved relative, Zane returns in the fall to the spots where he and his grandfather used to hunt together. Sadly, some of those old coverts are now no longer huntable—farms have gone under and "newcomers" have begun to trickle in, seeking pastoral charm, which is all the town really has to offer, but not the pastoralism that includes guns and hunting. For Zane, and for many others like him, this loss evokes almost exactly the same response one would have to someone vandalizing the gravesite of a loved one.

Few of the hunters I interviewed were as rooted as Zane. Most, like Keith Jones and Andy Felter, live far away from the sorts of memories that are refreshed on a daily basis for Zane as he drives to and from work over roads that are indelibly etched in his mind and which divide the woods and fields he has walked in all seasons of the year for virtually all of his life. Or they are like Rob, able to return a few times a year to familiar hunting spots to renew memories and maintain a continuity in their lives. These periodic returns to hallowed ground are all the more intense precisely because they are not routine. No doubt these travelers in time idealize their childhoods, as we all do, but idealization is not what drives them. Renewal of family ties and ties to the land are what keeps some hunters active (Andy and Rob) and what enrich the hunting for those for whom, like Keith and Zane, it lies at the center of their experience.

Most hunters have at least a few coverts to which they return year after year, even if there are other places where the hunting might be better. Hunting is part reverie; but hunting not only allows hunters to relive their own pasts, it can also transport them to an imaginary past, to a time they think of as a simpler way of life. It is a way of life they wish they had known directly. Karen DeFazio, the special education teacher and mother of five of whom I wrote in the preceding chapter, is no farm girl. She doesn't even garden. The daughter of Italian immigrants, she grew up in Boston. Now, she and her husband live in a bedroom community minutes from the freeway her husband takes to and from work. Time is at a premium for them both, but she and her husband manage to find the time to hunt for pheasants, turkey, and deer. Given conflicting work schedules, each hunts alone as often as they hunt together.

Karen's large family (the typical couple in the United States today has no more than two children) suggests a traditionalist inclination, despite her advanced degrees and her career. In fact, she does revere the past. Indeed, she said that if she were starting college all over again, she almost certainly would be a history major and would love to teach history. Her interest in the past is less antiquarian than it is her way of criticizing features of modern society with which she is uncomfortable. We've gotten too settled, too civilized, she said. "We expect things to be handed to us all prepackaged and dainty. . . . I've become the type of person that walks around in cemeteries, any cemetery, and just looks at stones and thinks back. . . . I get a feeling of being in touch. My kids say I'm getting historical. . . . I would have loved to have been born 150 years ago. I think I was born too late. *I would have been a great pioneer woman*" (emphasis added).

Karen likes the control over her own life that hunting embodies—she is engaged in an activity for which she is responsible at every stage, from beginning to end. There is no mediation, no division of labor (though there can be cooperation and sharing); it's just the hunter and nature in close engagement, much as she imagines it was when the country was still young. This direct engagement also carries with it the satisfaction of knowing, as Karen put it, "that I could feed my family, all five of the children. . . . That gives me a real sense of satisfaction, knowing that I know how to take care of things." Living within twenty miles of the Boston metropolitan area, a trim, athletic mother of five and a teacher imagines herself as a pioneer. Hunting gives substance and a sense of verisimilitude to her musings about the past. The past comes alive for her, and she feels connected to it, able to empathize, to imagine what the real pioneers must have had to endure as well as what they had to celebrate. Despite the fact that she and her husband are thoroughly enmeshed in the economy, she takes comfort in thinking she and her family could survive without all the creature comforts and shelves lined with food that contemporary Americans take for granted.

Self-Reliance

In the early decades of the nineteenth century, just as the Industrial Revolution in America was beginning in earnest, factory owners had difficulty keeping a stable workforce. Land was abundant and fish and game were still plentiful. This meant workers were not completely beholden to the factory owners. Since they could squeak by on what they could catch, shoot, and trap, as well as what they could grow, they were far harder to discipline than their English and European cousins. People moved in and out of the labor force, sometimes involuntarily and sometimes voluntarily. They had, as it were, one foot in the modern world of industry and the other in the ancient world of agrarian and subsistence economy.

The Osgoods, Paul Julian, and several other hunters, in all roughly 20 percent of those I interviewed, replicate at least certain features of this pattern. Living close to the land, if not exactly off the land, gives the Osgoods, just as it gives Paul Julian, a sense of security and self-assurance. For most of us, nowadays, our fate is not in our own hands. Decisions in some remote boardroom can radically alter our affairs in the blink of an eye. But people like the Osgoods and the Julians, while they are not impervious to market forces, are to some notable degree insulated from them. They are in, but not fully of, our consumer society. Among those I interviewed, they are unusual only in the degree to which they have insulated themselves. Many hunters understood their hunting precisely in these terms, even though, as with Karen DeFazio, their self-reliance was more metaphoric than actual. Needing this security can be practical, as it surely was for Paul Julian, but it need not be at all practical, as Rod Granger's situation makes clear.

Rod Granger is also a general contractor. He grew up in one of Massachusetts's small cities, and for most of his adult life, he lived in and around urban areas because that is where work was most plentiful. In the late 1980s he bought land in a small town in north central Massachusetts. The town is only an hour's drive west of the Boston metropolitan area, but once you get off the main highway and the handful of franchise fast food and convenience stores that mark the town center, you could easily imagine yourself a million miles from a major urban center. The real estate boom in Massachusetts in the early '80s made Rod wealthy. Unlike some other developers who thought the boom would last forever, Rod invested his windfall in the stock market and pared back his construction business so that he would not have to carry any debt. With understatement, Rod simply said, "The Massachusetts miracle treated me pretty well." The house he built for himself made the degree of understatement dramatically clear. Partially shielded by trees from the road, the house loomed up as I proceeded down the drive. A detached, scaled-down house for his two young daughters is separated from the main house by a full-sized, in-ground swimming pool. He said the children's house was a "playhouse," but it had its own plumbing, electricity, kitchen, and bathroom. By any standards, it would make a nice apartment for a single person, at least a short single person. An expansive lawn surrounded the house. Along the edges of the lawn was one of the things Rod had moved to the country for: an elaborate and well-kept, raised-bed vegetable garden. Just as his house might well be the feature of a home builder magazine, his garden was the sort that one normally sees only in gardening magazines. He was proud of the garden and gave me a detailed account of how he had mulched and composted and built the soil up to its present fecundity. I looked in vain for a weed.

As I cast my eye across the beds, I spied a Have-a-Heart trap.* Since we were about to begin talking about hunting and killing animals, I couldn't refuse the bait: "What's the Have-a-Heart for?" I asked. "Damn raccoons will eat everything. Right now they're after my arugula so I put the trap over there." "Why not just shoot them?" The question took him by surprise. Clearly, he'd never contemplated shooting the raccoons with whom he was unhappily sharing his garden. Rod thought a moment and said, a bit self-consciously, that the raccoons were too much like pets, and besides, he believed in killing only those animals he intended to eat and he had no appetite for raccoon. (A few minutes later, when he was showing me the turkey chicks and pigs he was raising, he said that he had shot a fox the previous year as it was running off with one of his young turkeys. I refrained from asking him if he ate the fox. I discuss this sort of inconsistency in a later chapter.)

Though Rod did not come from a farm background and indicated absolutely no desire to farm for a living, he shares what Muth and Jamison

---

*A Have-a-Heart trap is a wire mesh rectangular "cage" with doors that can be propped open at each end. These doors are linked to a plate in the middle of the cage on which bait is placed. When the animal enters the cage to take the bait, the doors fore and aft drop down. The animal is unharmed and can then be dispatched or, more likely, transported some distance away, even though relocating wild animals is illegal.

(1996) aptly call a "rural epistemology" with those who have actually farmed: he celebrates coaxing food from the earth. The harvest, with its rich evocations, was life affirming and gave Rod's life a meaning far deeper than the meaning he could extract from being a contractor or watching his portfolio grow. Hunting, like raising turkeys and pigs for the table, is simply part of the harvest. Of course, hunting is different from pulling up carrots or slaughtering a pig. In some ways, though, slaughtering an animal you've fed daily and watched grow is harder than shooting an animal you "know" only abstractly. Rod made plain his feeling that gardening, raising animals for the table, and hunting were of a piece. It is what humans do. Take any element out of this whole and things quickly get screwed up. "People forget where they come from and what, in the end, is finally important—and that's when the trouble begins," he told me.

His fortunate economic position struck Rod, as hard times struck Paul, as luck of the draw. Rod made no bones about his luck—he was in the right place at the right time. By contrast, the fruits of a garden or the success of the hunt were not simply the result of luck. Rod, Paul Julian, and the Osgoods had gardens that grew because they mulched, hoed, and turned the soil. Venison and pheasants were in their freezers because they had mastered woodcraft and marksmanship and had learned how to hunt. They were living by their wits, not by fitting in to the vagaries of the real estate market or the fickleness of interest rates. Their competence as hunters gives them, despite all the differences between them, a sturdy sense of dignity and self-esteem that economic successes or woe does not change. In the same way, what might appear as feckless in Zane's or Jack Wysocki's reluctance to get jobs and settle down is really something different: the degrees, jobs, and income by which most of us measure our self-worth simply do not inspire these young men. Their self-esteem is rooted in outdoor skills in general and in hunting in particular. Hunting well, which does not mean always coming home with game, confirms their self-worth and earns them the respect of their peers and the adults in their world.

Several of the hunters from urban areas find hunting compelling because it helps them, as Peter Carbone put it, "reconnect with a simple lifestyle." Peter is one of the more affluent of the hunters I interviewed. He lives in the state's second largest city in a neighborhood where the city's professionals and business people reside. His house is large but in a neighborhood where all the houses are large; it is not imposing in the way Rod's is. His father was a professional in town, and his two brothers are also professionals, one a lawyer, the other a physician. After college, Peter gave serious thought to becoming a lawyer but hesitated. To fill time and enjoy his youth, he headed to Miami, where he took a job as a private investigator. The experience did not increase his respect for lawyers. He returned to Massachusetts "jaundiced and burnt out" and not particularly interested in working with people. Luckily, he had always enjoyed working with wood, and breaking the middle-class rules, he became a carpenter and, soon thereafter, a builder of custom homes. His love is finish work, coaxing character and beauty from the wood. All kinds of shortcuts and efficiencies can be incorporated into the framing of a building, but the finish work is still heavily indebted to the craft traditions that date back centuries. Modern tools make the work easier and faster, to be sure, but the standards by which such work is judged were established long ago and many of the techniques are similarly unchanged. Given his embrace of traditions embedded in his craft, I was not surprised that, when I asked Peter which of the three weapons commonly used in hunting (shotgun, black powder, or bow)* he would choose if he had to choose only one, he didn't hesitate: the bow. His reasoning was straightforward. He took up archery initially just to extend the season and do some scouting before the two-week shotgun season began. But he quickly discovered that he preferred the total experience of bow hunting over shotgunning, despite the fact that, after ten years, he has yet to shoot a deer with a bow. "I see more game. During bow season, the woods are quiet and the animals are less spooky so I see all kinds of things that I don't see during the gun season."

This patient devotion, the willingness to sit absolutely still, hidden by camouflage, watching, listening intently, and waiting for a split-second opportunity to get an unobstructed shot at close range, makes Peter feel as though, at least for those hours afield, he is transported back to an earlier way of life, a way of life that appears more honest, more straightforward. Just as traditional woodworking methods inform his work, bow hunting in particular and hunting in general add to his sense of mastery and competence, without which he'd be just another builder, another faceless member of the middle class.

It is impossible to infer from a person's occupational status or income or education the degree to which he or she is psychologically committed to

*Massachusetts does not allow the use of rifles for deer hunting and has, like most states, special segments of the hunting season for bow and primitive (black powder) weapons.

the culture of getting and spending. Some, like Karen or Peter, can move freely between rural and urban rhythms and mind-sets. Others, like Paul Julian, the Osgoods, or Rod Granger, do what they have to do in the formal economy, but their hearts are really closely linked to the land. And then there are the young men like Zane and Jack, guys who are trying to remain as aloof as they can from contemporary society. Some would uncharitably say they are refusing to grow up and accept adult roles and responsibilities. There may be a measure of truth to this, but that is only a small piece of the puzzle. The far larger pieces involve a largely inarticulate discomfort with the exactions of modern society. They, could they choose, would opt for participating in a society more nearly like ours was seventy-five or a hundred years or so ago. Affluent or poor, these people hunt out of a need to remain connected to the rhythms of a former time and to confirm their self-worth in terms radically different from those established by the marketplace.

Not everyone was as self-consciously reflective about the ways hunting provides a sense of continuity and of keeping faith with the past as the men and women I have discussed thus far. But there were only a handful of people whom I interviewed who did not acknowledge at least some degree of fondness for or curiosity about what the social historian Peter Laslett has called "the world we've lost." To be sure, this desire cannot be taken literally. Karen DeFazio was no more willing to give up antibiotics, to mention only one now indispensable modern amenity, in order to become a pioneer than Jack would have traded his automobile or his power boat for a six-shooter and a ten-gallon hat. Keith Jones knew full well that the life he was living in the city, and even more the life that his city-bred children would lead, was far more stimulating and rewarding than life on the farm. That is, these men and women are not stuck in the past, throwbacks hanging on for dear life while the world passes them by. Though some, like Zane and the Osgoods, have been remarkably able to maintain a distance from many features of contemporary life, most are fully engaged in the society as it is. Indeed, people like Rod, Peter, and Karen have done very well, some by the sweat of their brow in construction and others by virtue of education and acquiring highly refined skills. Others have not prospered—Lady Luck has not favored Paul Julian, and Zane and Jack are not likely to achieve even middle-class status. For all their differences, though, they are drawn to the past, to a time when people are thought to have had more control over their daily lives; where their knowledge could be directly tested and unambiguously confirmed;

where broad competency, not narrow specialization, was the measure of a worthy person. The past that valued self-reliance and self-sufficiency is a world they prefer to the world in which they live.

In this sense, we might think of hunters, at least a clear majority of those I interviewed, as historical preservationists. Unlike the professionals who go by this designation, what hunters are preserving are not buildings or artifacts—they are preserving versions of an imagined past. It is all too easy to dismiss this form of preservation, much as hunters dismiss representations of nature with which they disagree as the "Bambi syndrome." In simplified form, hunters see themselves keeping alive the ideals of self-sufficiency and self-reliance, ideals that they feel are being eroded, if not entirely obliterated, by the currents of modern society. Though none would starve without hunting, virtually everyone I interviewed took pride in being able to put food on the table in a direct and unmediated way, just as our ancestors once did. Critics of hunting miss this point entirely, mistaking the absence of a literal dependency on hunting for subsistence for the absence of the need for game. Hunters, at least the vast majority of those who hunt, need to put game on the table, not to forestall starvation, but to reaffirm the values of self-sufficiency and self-reliance. Crucially, this reaffirmation is not a matter of lip service. Putting game on the table is to *act* self-reliantly. It scarcely stretches things to say that the consumption of game makes one self-sufficient in the same way that the communion wafer and wine bring one into the presence of Christ.

Hunters speak disdainfully of people who have no clue about how the basic necessities of life might be obtained were supermarkets to close. They pride themselves on the practical skills they have acquired and continue to hone by virtue of hunting. To be sure, this is tricky business. I am not suggesting that the practical skills that most hunters possess would actually see them through some shipwreck-on-a-desert-island scenario, though a couple of the people I interviewed would fare pretty well even in the most arduous and demanding circumstances (I think of Paul Julian here, or young Jack Wysocki, or Karen DeFazio). What I am pointing to is the *idea* or *myth* of self-reliance and self-sufficiency that hunting sustains. Rich or poor, male or female, young or old, with only a couple of exceptions that I will examine shortly, hunters expressed their admiration for the resourcefulness and stoicism of our forebears. Some used the language of manliness in this expression, but it was not about machismo so much as about independence—standing on one's own two feet.

90

Karen Defazio was as emphatic about this as anyone I interviewed. Hunting was a way for her to demonstrate to herself that she could take care of herself and her loved ones with her own hands—if need be. I asked her if she thought about self-sufficiency in connection with hunting. She replied, "Oh yes, very much so. I know without a doubt that I could [make out]. Even with five children that would be no problem at all. And that does give me satisfaction. That's why I've already spoken with the children about killing, and I've had them watch while I pluck and dress a bird on the back porch. It's not about the kill."

It's not about saving money or having no alternative sources of nutrition either. And, crucially, it is also not about "sport" or "fun," though as I show in the next chapter, fun is certainly part of the story. It is about establishing one's self as a person capable of enduring stress, privation, and exertion in order to live. It is about being responsible for one's own fate. To hunt is to symbolically reenact living off the land, subsisting on what you can glean by hard work and the wits God gave you.

When Emerson and Thoreau celebrated the virtue of self-reliance and made it not only a character trait to be cultivated but also the bedrock of a free society, they were regarded as radicals. Indeed, they were. Each in his own way urged his fellow citizens to break with the past, to be suspicious of received wisdom and the dictates of custom, and above all, not to be a slave to fashion. Though the people I interviewed had a narrower, and thus less radical, notion of self-reliance in mind, they nevertheless were operating within this tradition. It is as if they were saying, "the ability to go it alone, even though I may choose not to do so, keeps me free."

For the most part, of course, this claim has to be understood metaphorically: we are not dealing here with men and women about to turn their backs on jobs, mortgages, and creature comforts in order to homestead in the boondocks. Hunting sustains the illusion that they could do so. People recognize that they are depending upon an illusion, but that does not diminish the power of the idea one bit. Karl Granitsch, the fourth teacher I interviewed, exemplified this as well as anyone. He grew up on the outskirts of Worcester in the late 1930s and early 1940s. He never really hunted as a teenager, though he had a .22 rifle and a shotgun by the time he was in high school. He'd go out with friends and shoot at tin cans, and as he put it, he would "spray a lot of useless lead in the air" when he coaxed an odd pheasant or grouse into flight. After high school, he served a hitch in Korea and used the GI Bill to pay for his college education. He has taught in public schools in southeastern Massachusetts ever since. The town in which he has lived has long been too built up to afford him access to hunting opportunities, and the areas familiar to him from childhood are also now subdivisions. There is no hallowed ground to which he can return. Since he cannot take vacations in the fall, hunting would be precluded but for the fact that he belongs to a club that leases a couple of hundred acres of some of the last remaining open land in private hands in the area and releases commercially raised pheasants for members to shoot. Though the club is, he assured me, not a fancy one, by which he meant that it wasn't exclusive, it affords members some approximation of an uncontrived hunt. This appears to be enough to sustain Karl's sense of self-sufficiency. He and his cherished Brittany spaniel spend several hours each weekend at the club, and he is active in club affairs, including helping to arrange for the yearly game dinners the club puts on for members and their guests. The birds he brings home, even though they are pen-raised and released only hours before he and his dog hunt them up, enable him to enact self-reliance and self-sufficiency.*

Karl dreams of going to Alaska when he retires; if his health holds out, he is likely to do so. Several members of the club to which he belongs have jobs that allow flexibility in scheduling vacations, which teachers do not enjoy, and they have gone to Alaska to fish for salmon and hunt for moose, elk, and bear. His envy of their freedom was palpable. Meanwhile, he contents himself with outings at the club and the tastes of game presented at the game dinner that are killed in remote places by others less tied down than he. When I told Karl that the day before, I had interviewed a young woman who lived in the Berkshires and who hunted nearly every day of the season, mostly by walking out the back door and into the woods and fields surrounding her house, he shook his head as if to wonder how he could have been so foolish as to take a job in such a densely settled part of the state. Ironically, the woman to whom I was referring, Elaine Stebbins, the taxidermist now living in the Berkshires, grew up in the town next to the one Karl lives in.

---

*Hunting, of course, is not the only way people acted out self-sufficiency and self-reliance. I have already noted that at least half the people I interviewed maintained substantial vegetable gardens, Karl among them. Many were also active do-it-yourselfers, competent with home and auto repairs, and a few were very accomplished working with wood or metal. When I pulled into Karl's drive in the late afternoon, he hailed me from the roof of his modest, forty-year-old, ranch-style house. He was getting things ready for the new roof he was about to put on the house.

The Osgoods, of course, also thoroughly fit this pattern—indeed, of all the people I interviewed, they come closest to living in the way that most hunters can only dream of. While an economic calamity on the scale of the Great Depression would wreak havoc with the Osgoods, their lives would be far less disrupted than those of Karl Granitsch or the builder, Peter Carbone. I raise the prospect of a sharp downturn in the economy because many of the people I interviewed indicated that their ability to hunt gave them some cushion, something to fall back on, should hard times once more befall our society. Some were more gloomy about the prospects for society than others, but no one dismissed the possibility of hard times returning. Peter Carbone, for example, thought that a depression was unlikely but nonetheless expressed satisfaction knowing that he and his family would have meat, no matter what befell them. He quickly added, with a wry smile though, that this was pure fantasy: "There are too many people and not nearly enough animals to go around. So I really don't think we could go it alone. But the idea, knowing I know how to kill, butcher, and prepare animals for food, is reassuring."

Even in the comparatively good times of recent years, some people wind up in difficulty. Five of the men I interviewed were enduring hard times. I have already introduced Paul Julian, the contractor who had not had work for nearly a year, and Robert Swipe, who had had the rug pulled out from under him by his employer. Calvin Jones was working when I met him, but as we talked, it became clear that his work was rarely steady. A carpenter nearing his fortieth year, he combined working for contractors with small jobs he lined up on his own. But he said there were long stretches when he'd be unemployed. He was unmarried, so that made it easier to handle being out of work. Filling the freezer with game matters to Cal because there are times when money is so short that even the cheapest hamburger in the supermarket is a stretch. Most years though, he doesn't have to worry—between a deer and a variety of small game, Cal has meat to tide him over the lean months of winter. This not only takes the edge off unemployment, it also gives Cal a sense of control—he is not a captive of the contractors who hire and lay him off. It is not clear whether Cal even wants to work year-round. In a sense, he is a grown-up version of Zane or Jack, who are themselves the modern equivalents of the nineteenth-century men who moved in and out of the factory system as mood or the call of the wild dictated. What Cal strives to arrange are periods of unemployment that coincide with the fall and winter hunting seasons. This ideal arrangement

93

doesn't always work, but in the fall before I met Cal it had. That fall he got two deer, remarkably similar bucks, and then fell into a steady stream of jobs that had given him more cash than usual. He is using some of this unusual supply of cash to pay for an impressive taxidermy job on the two bucks. So proud was he of his feat that he insisted that we drive to the taxidermy shop where his deer were on display, waiting until he could make the final payment on the job.

Others were less adapted to the periodic loss of income. Greg Gougeon is a man in his early forties, father of three children. After having been unemployed for well over a year, he had just found work, a night job, a couple of weeks before I interviewed him. He looked rumpled and apologized for being "out of it": he hadn't adjusted to his new schedule. His wife worked, so that had taken some of pressure off—he didn't have to sell his boat, for example—but the strains were evident: a house in need of paint, screen doors without screens, that sort of thing. He had gotten three deer over the course of his being out of work, and he had availed himself of one deer that had been killed by an automobile. Unlike Calvin, Greg did not welcome the loss of work, no matter the timing. He did not go into details, but it was clear from what he said that it had strained his relationship with his wife and put everyone on edge. When I asked him how things would have been had he not hunted and fished, he just shook his head, not wanting to contemplate what things would have been like had he not been able to contribute to the family in the ancient way. Without any trace of irony or hyperbole, Greg said that hunting allowed him to "go back in time." What he meant by this, unlike some of the others I've been discussing who yearned for the past, was less a desire to live in preindustrial times than a sense of resilience and competence that enabled him to weather adversity. Without hunting and fishing, he would have felt completely helpless.

Roland Morreau also had had the rug pulled out from under him. For years he had been the manager of the service department of a family-owned auto dealership in south central Massachusetts. One by one, the owner's children went off to college, and when it became clear that none were interested in taking over the business, the owner decided to close up and retire. He did not even try to sell the business and instead auctioned off his inventory and the equipment. Roland was more than a little bitter about this because, had the business been sold, he at least would have had a shot at keeping his job with the new owner. Since then, Roland has held a succession of sales and delivery jobs, all of which he found too physically demanding for a man nearing sixty (who looks considerably younger and appears to be in good shape) or required long periods of time away from home. He has always fished and hunted and now does so more than ever. I entered his modest but carefully maintained home through the den and was greeted by a spectacular deer mount—a perfectly symmetrical ten-point rack of antlers on the head of an animal that dressed out at two hundred pounds—a very large deer by southern New England standards. There was also a bearskin rug in front of the glassed-in fireplace as well as the head of another bear on the wall opposite the deer mount.

Roland is completely at home in the woods. He told me that at least once or twice a year friends will call him to ask his assistance in tracking a deer they've wounded. Without a trace of braggadocio, he said, "I've never lost one." His friends have nicknamed him "the bloodhound," a moniker he carries proudly. Roland is a short, compact man, and he said mirthfully that this helped him because his eyes were closer to the ground than most peoples'. He hunts both small and big game, but his passion is for deer. Pheasant hunting is fun, a time to get out with friends, engage in banter and teasing, and bring back a bird or two. By contrast, deer hunting is serious business. A white-tailed deer hunt demands every finely honed sense and every bit of woodcraft a man or woman is capable of possessing. Deer hunting also requires putting up with discomfort, physical stress, and frustration, all the while remaining patient and clear-headed. It is the ultimate challenge, and to be successful at it is as clear an indication of self-possession and strength of character as anything Roland can think of. "Sure," he said to me, "anyone can get lucky and get a deer once or even twice. But no one can get a deer year in and year out without being really good in the woods. They have to really know what they're doing, and they have to be tough."

By "tough" Roland didn't mean simply physically strong or manly—though he hunted only with men, he claimed to know several women who hunted and whom he regarded as equals—what he meant was the combination of mental and physical toughness we associate with agrarians and others, like loggers, who make their living from the land. They have to be able to endure privation, to get by elementally, to improvise—in a word, they need to be able to do what it takes to survive. Though he did not use the word "self-reliance," there is no doubt that this is precisely what he was talking about. People should be resourceful enough to be able to take care of themselves. Hunting, at least when it is done seriously, tests that ability exquisitely.

For people like Roland, whose capacity for wage earning has suddenly declined dramatically, being able to affirm self-sufficiency and self-reliance is no small matter. Of course, this affirmation is almost purely symbolic. Even when a person can bring home appreciable amounts of game, as Roland clearly has been able to do, there is no real net savings involved. By the time one adds up the costs involved in hunting and fishing and spread these costs out per pound of fish and game brought to the table, they become very expensive sources of protein. Even for a family like the Osgoods, in which several family members in any given year are likely to bag at least one deer and a number of birds and fish, the costs of licenses, guns, ammunition, to name only the most obvious outlays of cash required to hunt and fish, make store-bought chicken or hamburger cheap by comparison. But because we are talking about symbolism, these grubby calculations are beside the point. Hunters feel themselves connected, by virtue of their hunting, to the ideals and the practice of self-reliance and self-sufficiency. This is so as intensely for Robert, who hunts with little of a practical sort to show for his efforts, as it is for Roland or the Osgoods, who eat game regularly year-round.

Perhaps, in the fabled days of yore when things were presumably simpler and more straightforward, a person's mental representations of his or her life coincided rather fully with what that person actually did. Frankly, I doubt it. So far as I can tell, our capacity for fooling others is matched by our capacity to fool ourselves, and neither capacity seems of recent origin. Simplicity, anthropologists have convincingly shown, is like grass—it's always greener on the other side of the fence. Still, the ideas persist that things were somehow better and that people were more honest and trustworthy in the past. Many Americans hold fast to comforting misconceptions about how warm and nurturing family life once was, conveniently forgetting the many who, like Robert Swipe, lived in almshouses or orphanages or those who were raised in unsympathetic families to whom they were sent to earn their keep. The hunters in this study are, in this sense, no more caught up in romanticizing the past than anyone else. What makes them different is that, unlike most nonhunters, they enact their nostalgia through their hunting. This enactment imparts a sense of empirical substantiation to their images of former times, as if they were stepping back in time and, for the moment, living and acting as their forebears had lived and acted.

IV

HUNTING FOR SPORT

Support for sport hunting is declining, though this may be as much the result of the general public's misunderstanding as it is a reflection of a distaste for hunting. The public, by a wide margin, approves of hunting for food but does not approve of hunting "for sport." Two sorts of confusion are at work here. Most hunters do indeed hunt for meat, but not because their lives depend upon bringing home game. Game is highly prized as table fare, and most hunters make something of a fuss over the preparation and consumption of game. Relishing the flesh of the animal you kill is part of the sporting ethic, a way of honoring the animal whose life you have taken. When I asked hunters whether they hunted for sport or for meat, virtually all of them said they hunted for sport simply because they were not driven by the prospect of starvation. But, with only one exception, Russ Farina, they quickly added that they really enjoyed eating game. Meat was neither irrelevant nor trivial.

In his study of hunters and antihunters, Kellert (1978) drew a distinction among hunters who hunted primarily for meat; hunters who hunted primarily for sport, by which he meant the quest for trophies; and hunters who hunted to be immersed in nature. I encountered no one who could be regarded as a trophy hunter, someone who was preoccupied with killing a specimen of an exotic species or an unusually large game animal. To be sure, any hunter would be thrilled to see, much less to bag, a deer with an impressive set of antlers, but no one I interviewed had an impressive rack (or the equivalent measure in other game species) as his or her preeminent goal. Indeed, when the subject of trophy hunting came up, most expressed distaste for it. As I shall show, when the hunters I interviewed said they hunted for sport, they had something other than trophies in mind.

What appears to bother nonhunters is the mistaken notion that sport hunters kill for the fun of it. This is a confusion about the meaning of the term "sport." The idea of sport hunting was introduced, as I noted in chapter 1, in the decades following the Civil War in an effort to elevate hunting above mere slaughter and to instill a conservation sensibility as well as a heightened sense of ethical responsibility the hunter owed wildlife. Sport in late-nineteenth-century America meant a gentlemanly undertaking bounded by rules of fair play and good sportsmanship (Guttmann 1978). The point was the love of the contest, not winning or losing. It is ironic that in a society as obsessed by sports as ours, the seriousness of sport gets lost when applied to hunting. It may not be stretching things too far to claim that, at its best, sport hunting represents a purer example of the ideal of sport than any of the more common sports. Sport hunters are much less likely to adopt the win-at-any-cost attitude that has come to infect other sports at all levels, from youth programs to the professionals.

Even when winning matters to the hunter, as it must have to the aboriginal or to the colonial hunter whose very existence turned on "winning," there is pleasure in a failed hunt. The activity is pleasurable in itself, even when no shot is taken. This is no doubt one of the reasons the Puritans were so conflicted about hunting: they needed the hides and meat, for immediate consumption as well as for trade, but they feared that those who hunted would become so absorbed by the activity that they would lose all self-restraint and piety. Our attitudes toward sport are no longer freighted by the stern admonitions to piety and self-restraint of the Puritans, but when it comes to hunting for sport, the general public thinks the animal being hunted is devalued. Nothing could be further from the truth. In order to begin to appreciate this, we need to reflect on what hunters mean when they say the hunt for sport.

As I listened to hunters describe their motivations, three distinct elements emerged: hunters talked of hunting being "fun," of being a form of recreation, and, finally, of its being a sport, by which they meant a contest in which the hunter matched wits with his quarry. An important dimension

of each of these motivations was the enjoyment hunters derived from being close to nature—observing wildlife, seeing the traces of long forgotten
farms and the ways nature had reclaimed the former fields and pastures,
and enjoying the spectacle nature afforded. Of course, these are not mutually exclusive motivations, even though there is tension between having
fun and being a good sport hunter. Each hunter balances fun, recreation,
and sport in his or her own fashion, and this gives hunting a range of mean-
ings, not only varying from one hunter to another but also varying over
time for each individual hunter.

For the Fun of It

Hunting is fun—if you enjoy the out-of-doors and getting off the beaten
path; if you enjoy mixing exertion with intense concentration; if you like
the suspension of workaday rhythms and codes of decorum; and if you like
your activity tinged by a measure of risk.* These tastes are not for everyone, and hunting is by no means the only activity that offers these de-
lights—whitewater kayaking, technical rock climbing, and serious mountain biking and downhill skiing come to mind as activities that share these
qualities with hunting. But these activities lack the element that is unique
to hunting—the kill. While many hunters engage in several strenuous outdoor activities, all acknowledge that hunting is different. Even if the desire
to bag an animal is not high, hunting is not just a walk in the woods. The
gun changes things.

Margaret Jacque, a divorced mother of two grown daughters, one of
whom hunts, works as a legal secretary in a small but busy office in a once
bustling, small industrial city in north central Massachusetts. As a young

---

*The risks entailed by most hunting were once quite large, less because hunter and hunted
could change places and more because hunters were not all that well trained, which meant
that accidents were common. But after more than five decades of concerted efforts by the federal government, state agencies, and hunters' clubs and organizations, hunting now is one of
the safest outdoor recreations. Accidental shootings are rare and most involve self-inflicted
injuries. In 2000, the latest year for which I have been able to get complete data, there were
nineteen fatalities attributed to hunting, one of which involved a hunter shooting a nonhunter by mistake. By comparison, in the 1930s, many states recorded annual fatalities far
larger than the total now for the nation as a whole. I explore this again, more fully, in the last
chapter, particularly the question of why hunters are thought to be a threat to the general
population, despite data that demonstrate that hunters have achieved a remarkable safety
record in recent decades.

person, she and her family did a lot of camping, mostly on Cape Cod, and she remembers a childhood filled with talk of the outdoors. Her father hunted but was not very serious about it and made no effort to involve Margaret or her siblings in hunting. But the man she married was an avid hunter, and when their two children were old enough to allow her time to herself, Margaret began tagging along, "just to see what it was like, and since he was so into it, I thought it would be a good thing for us to share. And I always liked getting out for hikes."

After a few of these outings, Margaret decided to take the hunter education course that all new hunters must complete in order to be licensed and to be able to carry a firearm. Her husband got her a shotgun, and they went to a local rod and gun club, where she learned to shoot skeet and trap. An energetic and outgoing woman, she enjoyed acquiring this new skill and, even more, enjoyed hunting pheasants with her husband and their dog, a mixed Lab and golden retriever. She also went deer hunting and duck hunting, but the long periods of sitting on a stand or in a blind were less appealing than the constant movement involved in upland bird hunting. Learning to "read" a cover, to be able to figure where pheasants are most likely hiding, and to "read" her dog—to understand how its movements, the action of its tail, and the way it carried its head signaled that a bird was nearby—were things that she described as thrilling. It mattered less to her that she shot a pheasant than that she had come to know enough to be in the right position for a good shot. Though she never became as good a wing shot as either her husband or her eldest daughter, whom she encouraged to get involved, she became good enough so that she never felt she had to prove anything and no one had to make excuses for her when she missed. That meant that she felt no pressure to improve or to become as serious about hunting as her husband was. "It was just plain fun to get out with the dog."

Even when her marriage began to fall apart, hunting remained something she and her husband continued to enjoy together. Indeed, Margaret recounted hunts that could have added to the strain in their already strained relationship but instead were among the rapidly declining opportunities for shared laughter. She laughed aloud when she recounted their last outing for ducks—a particularly nasty morning, the sort of morning only duck hunters appreciate. Snow was falling steadily, and it made it easier to see in the dark as they carried their canoe across a field to the river where they would set out decoys and wait for the ducks to come in; but it didn't add to their comfort in any other way. The hoped-for ducks never materialized that morning, and by the time they decided to call it a day, Margaret was freezing cold. Feet and hands numb, wet from helping to round up the decoys, she slipped getting out of the canoe. She and her gun landed unceremoniously in the mud and snow on the riverbank. Her eyes watered from laughter as she told me how her husband slipped and fell trying to assist her.

Despite the gravity that lies at the heart of the activity, hunting involves a lot of frivolity and slapstick. Funny things happen. Mostly, the humor is the sort that requires having been there or knowing the butt of the joke, but all the people with whom I spoke related at least one comic moment, often, as with Margaret's tale of mirthful woe, with the joke being on them. Encounters with electric fences and cow pies, sometimes at the same time, were common. Retelling these stories with companions helps enrich and embellish the hunt and relieves the tensions of a day spent "on alert."

Margaret has not hunted since she and her husband divorced. She has given thought to going out by herself, if only to get the dog out (she "got the dog" when they divided their belongings), but she is reticent about entering what is still a solidly male domain by herself. She'd like to go with her daughter, but like so many single mothers and single young adults, they both work long hours and never seem to have the same free time. Moreover, her daughter has just decided to apply for law school, so she will have even less free time than she has now for some years to come. When I asked Margaret if she realistically thought her hunting days are over, she said in a serious tone, "I hope not. I really enjoyed hunting and miss the fun of getting out on a crisp, fall morning with the dog."

Dick Board, had he had to stop hunting, would say the same thing. A tall, wiry man who had just finished his first year of retirement after thirty years working for a public utility company, Dick hunted three or four times a week during the first fall bird season of his retirement. Among his friends, he is the first to have retired, so, though he prefers to hunt with one other person, he found himself hunting alone quite a bit—well, not exactly alone. His German shorthaired pointer is his constant companion, at home and in the field. Years ago, Dick was also a pretty serious deer hunter, but as areas to hunt began to shrink, the areas he hunted got crowded, and he began to get concerned for his safety. He quit hunting deer for good when he encountered another hunter in the woods who was quite drunk. That was enough for him. Bird hunters, at least in his experience, are different.

Drinking is not part of the lore of bird hunters. Besides, Dick noted, there's far more action involved in bird hunting. "In a good *season,*" he averred, "I'd see three or four or, at the most, a half-dozen deer. Hell, on a good *day,* I might see a dozen birds, not that I'd have shots at a dozen, but at least I'm seeing plenty of game. And that's not counting the rabbits we scare up but don't shoot."*

It's not just that Dick sees more game while bird hunting—he sees birds because he has trained his dog to hunt in close coordination with him so that when they come upon the scent of a pheasant, the dog slows down; and when it gets to within five or ten yards or so of the bird, the dog freezes. This allows Dick to get close enough to see the bird clearly when it flushes. Everyone who has witnessed this kind of dog work understands how exciting those few moments are. I have taken people who are opposed to hunting out with me when I take my dogs out on training runs, and even they cannot help but be excited when a big, brightly colored cock pheasant rises, cackling indignantly, from a hedge row or the edge of a corn field. It is an impressive show.

Dick enjoys eating pheasants, but that's not what compels him to hunt; it's also not the shooting—he and his wife shoot revolvers and rifles in low-key local competitions, and this gives him all the shooting opportunities he could wish for. Dick hunts for the pleasure of seeing his dog work and for the rush he feels every time a bird flushes over one of his dog's points. And after the excitement is over, there is the calmer satisfaction, an afterglow one might say, of having seen the hours and hours spent with the dog paying off. Hunter and dog are not just companions, they are a team. Dick loves to hunt, but he made it clear that if something happened such that he could no longer have a dog, he would stop hunting. "Might as well just go for a walk," was how he put it to me.

Another bird hunter, Mark West, understands Dick's feelings implicitly. Mark grew up in the Connecticut Valley of Massachusetts when much of the valley was still farmland and early successional uplands. Beginning in the 1950s, when Mark was a boy, farms began to fold. Some quickly became house lots, but many lay fallow, awaiting surges in the business

---

*Most bird hunters who hunt with trained pointing dogs will not shoot rabbits while hunting birds. The consensus is that shooting rabbits over bird dogs will weaken the dog's desire to point, that is, to freeze upon scenting a bird nearby and remain motionless until the hunter either releases the dog to flush the bird or walks in and flushes the bird himself.

cycle that would transform the hay- and cornfields into prime real estate. In New England, fallow fields quickly begin to revert to forest. First wild raspberries appear; then in the wetter places alder show up. Birches and aspen and cherry and maple saplings are not far behind, and almost before you know it, prime ruffed grouse and woodcock habitat arises. On the farms that remained active, the cornfields, bordered by swale and scruffy, overgrown drainage ditches, made for excellent pheasant habitat. Mark grew up in what was very likely the heyday of upland bird hunting in Massachusetts.*

Mark's father did not hunt, though he did not discourage Mark's interest in hunting, so Mark began hunting with his young friends. At first they mostly hunted for squirrels and rabbits, without much luck, as he remembers it. For some reason—he thinks it might have been a combination of his father's lack of interest and his lack of respect for the adults he knew who hunted deer—Mark never gave deer hunting a thought. His first serious hunting, which he defined as when he began to spend a lot of time reading and thinking about hunting, was for birds. He has hunted waterfowl seriously but in recent years has concentrated more and more of his hunting time on grouse and woodcock.

His dog of choice is the Brittany spaniel, and for years he has been perfecting a line of dogs that has achieved a reputation among local bird hunters for the ease with which they accept training and for their ability to find and point birds. At any moment in time, Mark has three or four adult dogs and normally raises two litters of pups a year. Mark works out of his house, doing custom machining for clients all across the country, so it is not too hard to keep this modest but demanding kennel operation afloat. During the hunting season he works nights and Sundays (hunting is not allowed in Massachusetts on Sundays) so that he can get out with his dogs during the day as much as possible. "Unfortunately," he remarked ruefully,

---

*Since the early 1960s, the decline of farming has been precipitous, in Massachusetts as well as across the rest of New England. For a while, this decline produced prime wildlife habitat in those places that did not immediately become subdivisions or shopping malls or the now ubiquitous storage facilities, the most recent testament to our love of mobility. Paradoxically, this has meant that Massachusetts is now more forested than it was a hundred years ago. But the forest is aging, in many places now almost sixty to seventy years old. Such mature forest may be good for all sorts of creatures and all sorts of aesthetic benefits, but most game species do not thrive in homogeneously aging forests. Wildlife, and especially game birds like grouse and woodcock and pheasant as well as deer, need dense stands of young saplings and the brush and berries and thick swales that first appear in abandoned fields or in the wake of logging.

"business has been too good in the fall recently," so he has been forced to cut back on his hunting time.

Like Dick, Mark and his wife (their children are all grown and on their own now) eat the birds he brings home, and they also share them with friends, a common practice among the hunters I interviewed. But, again like Dick, Mark doesn't hunt for the food. He hunts because it is fun. The dogs, each with its own quirks, strengths, and weaknesses, are entertaining, even when they are not on the trail of a bird. Though they can also be exasperating—older, experienced dogs can suddenly get headstrong and decide that they no longer have to do things the way they were trained to do; young dogs can push at the limits as if to see how much they can get away with—the variety itself, from one dog to the next and from one day to the next with the same dog, adds a dimension of unpredictability to an outing that adds to the stimulation and pleasure of the hunt.

It is probably not an accident that it was the hunters who particularly loved bird hunting and working with dogs who emphasized the entertainment value of hunting more than the hunters who may have hunted birds but considered themselves deer hunters.* This may be because the dogs are so all-consuming and so rich a source of pleasure, quite apart from their contribution to the hunting experience. It may also be a consequence of the lower level of moral seriousness we, hunters and nonhunters alike, attach to birds as compared to deer. Killing a deer, by all accounts, is a far more overpowering experience than killing a pheasant or a grouse. It's not

---

*As luck would have it, no one among the people I interviewed concentrated on water-fowling. The lore and literature suggests that duck and goose hunters are different, in turn, from other bird hunters—as one might guess from Margaret's experience, not everyone is drawn to sitting in a freezing rain waiting for ducks to fly close enough to afford a shot. My few duck-hunting attempts may not be typical, but I have no reason to think that I alone have had such experiences. The first was when I was still in high school. In northern Minnesota in early November—it's *cold* in early November there—a friend and I put his canoe in a small lake known for its wild rice and thus for its flights of ducks. We paddled across the lake, thankful for the exertion that warmed us up, and slid back into the tall rushes to await the ducks. We could hear them before it was light enough to actually see them, but knowing we were surrounded by scads of ducks kept us from worrying much about the cold. As it grew light, a small flight of ducks came right for us. In our naïveté and excitement, neither of us had the presence of mind to coordinate the shooting. We both shot at the same time in the same direction, the force of which tipped the canoe over. I'll leave the rest to your imagination. I'm not a duck hunter. And I'm fully prepared to accept the proposition that duck hunters may have unique motives, masochism aside. Still, I do not think, when all is said, that the differences are so large as to distort the representation of hunting that I am providing here.

that deer hunters don't have fun or don't have plenty of very funny stories to relate about their hunting experiences; it is just that such talk was tempered in the interviews with hunters who concentrated on deer. Those deer hunters who also hunted birds talked about their bird hunting more casually, as if to say, "oh yah, I go out for birds, but not with the same intensity with which I hunt deer." Bird hunting is less symbolically loaded, and that may make it possible for hunters to talk about it as "just fun." This is a topic to which I must return when, in the next chapter, I consider hunting ethics and the moral burden hunters confront when they kill an animal.

The activity of hunting, whether for birds or deer, itself is fun in the same way that a youngster finds pleasure in sprinting in a new pair of sneakers—there's no stopwatch or finish line: there's just movement. Zane is someone who appreciates movement. As we talked, he was constantly shifting positions on the couch, not because either the couch or being interviewed was uncomfortable. He's just a bundle of energy. As I drove off after my interview with him, I wondered how in the world he had managed to sit through forty or so courses at the university. Though he does strenuous manual labor and comes home tired at the end of the day, one of the things he looks forward to is the physical activity that hunting involves. The activity is not repetitive in the way work tends to be, and one's whole body is involved, not just a few specific muscle groups. The tiredness one feels at the end of a day's hunt is completely different from the fatigue at the end of a workday. It is curiously satisfying, the aches and pains and stiffness a confirmation of having lived, not of having worked.

Almost all of the hunters talked about the physical demands of hunting and how much they enjoyed the exertion. I was about midway through the interviews when it dawned on me that I had yet to interview a single person who appeared out of shape. I was on a run one afternoon when that realization struck me. In the remaining interviews, I was more attentive to how the men and women looked and moved. Of the thirty-seven people I interviewed, only one was overweight, and only one had a pot belly. All the rest, young or old, male or female, were active, trim, and athletic-looking. Most said they jog or swim or bike to stay in shape. Five, all men, clearly worked out with weights, not to bulk up so much as to stay fit and strong enough to handle a heavy bow or carry an eight-pound shotgun at the ready hour after hour. The exceptions, two older men, had given up active hunting several years before my interview with them; one because his wife had become seriously ill, and he was uncomfortable leaving her alone for even a day; and the other because of his own health problems. This group of hunters would not provide much grist for the critics of hunting who portray hunters as pot-bellied, beer-guzzling sloths.*

Hunting contains the raw elements of play—variation, exertion, excitement, surprise—and almost everyone acknowledged that these were things about hunting they enjoyed. And for those, like Margaret, who had ceased or suspended hunting, these elements were what they reported missing the most. They didn't miss killing an animal, they missed the adventure and exhilaration of the hunt. Others, while acknowledging the fun involved in hunting, stressed other features of the hunt, aspects more nearly captured by the concepts of recreation and sport.

Recreational Hunters

Russ Farina does not like the taste of game. In fact, Russ is the only hunter I interviewed who did not make it a point to say how much they relish the taste of game. Russ gives what he kills to friends and neighbors who do like game. Russ hunts to get away from the demands and pressures of work. He owns a number of commercial and residential properties in the western part of the state and manages some properties for others as well. Keeping them rented and keeping tenants happy is not only a full-time job, it is a source of continual hassle, and in light of the vagaries of the local economy and the increasingly complex tangle of regulations, it is often nail-biting and exasperating. His holdings are large enough to ensure his financial comfort but not so large that he feels able to turn over the day-to-day operations to someone else. His life is filled, as he puts it, "with a good deal of static."

Hunting, especially bow hunting, represents about as complete a shifting

---

*On the subject of beer, I did not explicitly ask people about their drinking habits, though I did ask how they felt about drinking *while* hunting, which, not surprisingly, was uniformly and emphatically deplored. Drinking in the evening is another thing entirely, and though it no doubt means that some hunters go into the woods in the morning hung over, it is rare to encounter a hunter drinking in the woods. Most said they wouldn't even have a beer with lunch because, even though it wouldn't make them drunk, it would take the edge off and that could be the difference between making a good shot or not. In my notes, which I took after each interview to record the things that the tape recorder cannot, I noted that five of the men had made oblique references, either before or after the tape was running, to past problems with alcohol, and several made it clear that they were now "dry." One indicated that the incentive to go on the wagon was to become a more proficient hunter.

of gears as anything Russ can imagine. He mostly hunts alone, having had his fill of social interaction at work. When he was young (he is now in his early fifties) he hunted pheasants, but for the past several decades he has concentrated almost exclusively on hunting deer, which he hunts with bow, shotgun, and muzzle loader. As noted, he prefers bow hunting because it is most demanding. In recent years, Russ has begun to hunt turkey in the spring, which adds several weeks to his hunting opportunities. Both turkey and deer hunting require a lot of preseason scouting and then a lot of patient sitting. Russ does a lot of this sitting on his own land west of the Connecticut River and rarely travels out-of-state. Unlike many hunters with whom I talked, he has no interest in going to Montana or Alaska—all the arrangements and details strike him as too reminiscent of his work. What Russ appreciates most is the luxury of spontaneously canceling appointments when he's feeling stressed and getting into the woods. Hunting for him is pure escape.

Shedding his coat and tie and donning camo is like stepping through Alice's looking glass. Within minutes Russ is in another world, a world in which he is carefree. He is liberated from the world of adult responsibilities. Whereas other hunters may hunt in order to feel self-sufficient, such thoughts have not crossed Russ's mind. To be sure, as I pressed the point a bit, he did acknowledge that, especially while bow hunting, he sometimes has fleeting thoughts of Indians and a slight touch of envy for what he imagined was their uncomplicated life. But he had no desire to live closer to the land or to simplify his comfortable, albeit hectic, life. As long as he can arrange to get out and be alone, he is willing to put up with the grief that comes with his affluence.

But what's so special about hunting, if escape from responsibility is the goal? I asked him why the solitude of a long walk in the woods or lazily paddling a canoe on the flat water stretches of the Connecticut River would not do just as well? Russ thought about this for a moment before replying. "I do those things, or the equivalent, when the hunting season is closed, but it's not the same," he replied, stroking his chin as he spoke as if to coax more words from himself, words that would name the difference. Russ is no stranger to words, but the right words did not come. So I pressed on. "Is it the intensity and concentration that hunting requires that makes it so easy to forget everything else," I asked. "Yes, that's it," he said, and then continued. "When I'm on a walk, my mind wanders and inevitably drifts back to something that's been bugging me or something I ought to have done better. But when I'm hunting, I am focused. All I care about is figuring out if that slight noise I just heard is a deer coming into my stand. I am totally absorbed, in a completely different world."

Hunting is different from a walk: at any moment, after all, a life-and-death drama might unfold. Think of how rare the occasions are when a vague noise enlivens all of our senses and starts the adrenalin coursing to our muscles; something that goes bump in the night—an intruder? There simply are very few occasions when we are as alert to our surroundings as a hunter must be. Add the anticipation of a deer or a turkey coming into view, and it is easy to understand why Russ finds hunting so completely engrossing. As a hanging did for Samuel Johnson, hunting concentrates the mind. Everything gets pared down to the most elemental essentials of what you can hear, see, and smell and how steady you can be when the hoped-for shot comes along.

To be sure, no hunter can maintain such peak intensity for more than short intervals. There is room for daydreaming and wool gathering, even for dozing off when the listening and waiting begin to seem like watching paint dry. But even the dozing is different—it's as if only your vision is put on hold: your ears are still awake to that telltale snap of a twig or sudden cessation of a bird's song. In a flash, you are once again on full alert, daring not to move a muscle even though your nervous system is urging action. At those moments, which can dissolve into the comedy of discovering that the sound was made by a chipmunk scurrying over dry leaves, or the marvel of watching a fox heading home with a rabbit, there literally are no cares in the world. The anxieties of unemployment and the hassles of a successful businessman alike dissolve into thin air.

George Dunnerston is retired and would appear to have no cares from which to escape. He is vigorously active and in good health, as is his wife of nearly forty years. When I met him in mid-summer, he was well tanned from almost daily golfing and from working in his intricately laid-out flower garden that made his trim little house stand out like a jewel in the neighborhood of modest homes in a small town in the Berkshires. He began hunting as an adult, urged to try it out by business associates (he was an engineer who moved into management in a large manufacturing firm). He liked the experience of going out with friends and relaxing. In camp (a generic term that covers everything from a cramped camping trailer to an elaborate lodge), the edginess and competitiveness of men on their way up (or trying not to falter) was largely absent, and it was a welcome respite.

George has hunted for approximately twenty-five years, mostly for deer, and has killed two. (A classic mount of the head and antlers of the largest one adorns his parlor wall.) That low success rate, translated into business terms, would have gotten him summarily fired. Translated into golf, it would have made him the laughingstock of his circle. But hunting is different, at least for many hunters (I shall introduce some, shortly, for whom consistent success is important). Nobody keeps score.* George emphasized this when he said, "If I felt I had to get a deer to prove something to someone, I wouldn't enjoy it."

George still hunts with some of his former business associates. The sociability of gathering with old friends, maybe even maintaining a link, albeit slender, to the firm and what's new in the product lines he had helped develop over the years, draws him back each year. The timing of the season is just right, too. The golf courses have long since closed for the year, his precise flower beds have been mulched and pruned against the coming of winter, and it's too early for his annual wintering in Florida. Hunting fits precisely in the crease between the other things George does for recreation. It's refreshing to get away, to break up even the fluid rhythms of retirement, and to use one's limbs in different ways. It's also refreshing to suspend the ordinary rules of decorum—he can forgo shaving, let go of inhibitions, and indulge in food and drink that would in other settings be judged excessive. Suspension of such rules makes the distinction between work and recreation all the more sharp and makes the experience refreshing because of the contrast.

George is not nearly the serious hunter that Russ is, so the elements of pure fun are more pronounced in George's talk about his hunting than they are in Russ's, but if they ever met and talked about hunting, they would talk the same language. This would also be so were either Russ or

---

*In Europe and in the nineteenth and early twentieth centuries in the United States, the elite hunters did keep score of the amounts and kinds of game bagged: the more the better. That has changed dramatically, though as I shall show later, the change may be eroding as a result of the commercialization of hunting (private leases, pay-as-you kill arrangements, and the like). A personal anecdote may help here. A number of years ago, a good friend and fellow bird hunter and I were playfully teasing each other about the relative merits of our dogs and our respective abilities with a shotgun. It ended in a good-spirited, friendly wager: let's see which of us gets the most birds by the end of the season. By the end (my friend won the bet) we both agreed that it had been a terrible idea to introduce that element into our hunting. It didn't take all the fun out of it, but it did diminish the fun. Were keeping score a central feature of hunting, we would not have been nearly as avid as we were and have remained.

George to meet Tommy Worthington, one of the more consistently successful hunters I met. Tommy and his brother run a large auto repair shop in the Berkshires, and both are avid hunters. Their shared office has several impressive deer mounts on the walls, as well as several turkey fans and numerous pictures of one or both of them with fish and game proudly on display. They grew up in a hunting family, and as soon as they were old enough, they accompanied their father, uncles, and cousins on hunts near home as well as in neighboring New York State and at a family camp in Maine. Tommy's brother, who is a few years younger than Tommy, still hunts in all three states, frequently with family members. Tommy has been hunting less often in the past couple of years because he and his wife have three young children, and for the past two years he has been building a house, working on it after work and on weekends. When I interviewed him, he was only a couple of weeks away from finally moving in. He leaned back, anticipating the contentment to come, and said that with the house finished, he could at last get back to hunting more.

The fall before our meeting Tommy had failed to get a deer, the first time in a long time that had happened. "I just didn't have the time to do the preseason scouting, which I usually combine with grouse hunting, what with the house and all," he explained. And he felt obliged to forgo out-of-state hunting that year. But he made it clear that he wasn't disappointed about not getting a deer.

> When I was young, I really had to get a deer. I was trying hard to be as good a hunter as my dad and uncles and was out to prove something. But that ended long ago. I guess it was after the year I got eight deer. The pressure was off after that. I don't feel that pressure any more. It didn't break my heart not to get a deer last year. Now I hunt to get away, to enjoy myself and be with friends. I usually do get a deer, but that's because over the years I've learned a lot, and having lived around here all my life, I really know where to find game, so I have a real advantage over most hunters. But I don't need to kill something anymore. I don't need to prove anything any more.

Getting away was a recurrent theme in my interviews. For some, like Russ, hunting got him away from work. Tommy also valued getting away from the hassles of managing a business—employees who didn't show up for work, customers who never seemed satisfied, haggles with insurance companies, and mounting regulations of the hazardous materials he used in the garage all gave him good reason to get away now and then. But it was also clear that, as for several others with whom I spoke, getting away also meant getting away from the responsibilities of being a father and husband.

Tommy and his wife hunted together when they were first married, but when the kids came along, she put up her gun and bow. Though he said he hoped she would return to hunting once the kids were grown, his voice seemed to lack conviction. He was even more vague about introducing his kids (two daughters and a son) to hunting.* Though he was clearly grateful to his father for having encouraged and included him in the hunt, he seemed decidedly unenthusiastic about following in his father's footsteps. He didn't say so in so many words, but I was left with the distinct sense that the escape he sought would not be nearly as complete or as gratifying were he to have one or more of his kids in tow. He hunted in part to get away from nagging and impatient kids (which, interestingly, was how he described himself as a child) and feared that including them would ruin his experience. Though I can only speculate on this, it may be that Tommy had felt the need to prove himself so intensely as a young hunter because he had been made to feel more like a fifth wheel than a welcomed companion.

Few hunters shared Tommy's lack of enthusiasm for introducing their children to hunting. But several others made it clear that the time they spent hunting was time that they did not have to deal with the complexities of domestic life. While the demands of work were far more frequently cited as a source of frustration, especially when they cut short the time available for hunting, it was clear that for some, hunting was an escape from family responsibilities. Except as regards safety, deer hunting permits the hunter to be oblivious to the needs of others. You are in the woods, and even if you are hunting with a group, you are rarely within eyeshot of one another.** Some hunters carry two-way radios so that they can stay in touch, both for safety reasons as well as to alert the group to the

---

*I discuss fully in chapter 6 hunters' feelings about their children's learning to hunt.

**Upland bird hunting is different. If you are hunting with one or two other people, you are almost always in sight of one another, or at least in hailing distance. Because the flight of a bird is unpredictable, it is important to know where your companions are. Not only does this minimize accidents, it also means that if the bird flies in such a way as to offer no shot to one member of the party, the other might have a shot. Bird hunting with a friend or two is more social during the hunt. Parties of deer hunters do their socializing after the hunt.

whereabouts of a deer. Tommy had used a radio only once and didn't like it at all. It was cumbersome, but more importantly, it disturbed his peace and quiet. He felt as if he was "on call."

Fred Jenkins is the father of seven children, ranging in age from six weeks to ten years when we met at his large but sparely furnished home in a mill town in central Massachusetts that had seen far better days. Fred is a mason and bricklayer who works for a large construction company. The work is steady, though it slows down in the late fall, and as we have already seen with others in the building trades, he doesn't complain about having extra time when deer season rolls around. Though he makes a decent living, he has lots of mouths to feed. Even though this responsibility gave Fred a very practical reason for hunting, he still made it clear that he hunted for recreation as well as for food. Nature intrigued him, and he enjoyed watching the antics of chipmunks and the behavior of birds, marveling at the ways they all fit together into a whole. But he was also drawn to the traces of prior human activity. Being a stone mason, he was especially interested in the cellar holes and stone walls that are, along with the fall foliage, virtually trademarks of the New England landscape. Man's handiwork and nature's form a rare fit in the woods of New England, at least when all that's left are lichen-covered walls. "Boy, these stone walls amaze me. The physical labor that went into them just amazes me. And even though the trees have all grown in, they are still standing and they didn't even use any cement or anything. I'll be out there and study them and say 'this guy was an artist' or 'this guy didn't know what he was doing.' But good or bad, it sure was work."

Fred's work life is spent making the built world into right angles, defying natural forms. It is relaxing and even a bit reassuring for him to see the remnants of a built world that have, with time, become "naturalized." He put this in terms of control—when he enters the woods, he relinquishes control to nature. "When I'm around here [the house], I'm the boss. When I go hunting, nature's the boss. When I'm in the woods, nature calls the shots. You never know what to expect because everything is constantly changing. There are just so many variables when you go hunting." The anticipation of surprises, of new experiences from which one might learn something about nature or oneself, makes being in the woods both challenging and humbling. And the tension between nature "calling the (big) shots" and the hunter being able to call only the "little shots" yields an experience that is utterly distinct from the rest of his life.

Unlike some of the other hunters I interviewed, Fred does not feel at home in the woods. He carries a high-quality compass and the topographic maps of the area he's hunting, and even though it's a bit of a burden, he also carries a fairly elaborate first-aid kit with him. Neither he nor the few friends with whom he regularly hunts has ever had an accident; but his clear sense is that when you leave civilization, you enter a world of risk, and it pays to be prepared for whatever nature and the fates have in store. Fred hunts not so much to escape responsibility as to be refreshed by the contrast between the world in which he is responsible and in charge and the world in which he is at the mercy of nature.

Psychologists have long recognized that we humans need variety and the stimulation that comes with new experiences. We need to get out of our routines, our ruts, every now and then. Hunting clearly satisfies this need for the men and women I interviewed. Whether they were casual hunters who were really just walking in the woods with a gun, or serious hunters who spend many hours scouting or training dogs or practicing with their guns or bows; whether they were affluent, or nearly impoverished; whether they were single, or had large families to support; whether they were employed, unemployed, or retired, hunting represented a change of pace, a break in the routine, a chance to cultivate and refine skills not called upon or even much honored in the workaday world. In short, hunting is recreation. Hunting is also a challenge, and some hunters take this challenge more seriously than most. In addition to being fun and a form of recreation, hunting is also a sport.

Hunting as Sport

All but a handful of the hunters I interviewed could be called "avid." In the off season, which is most of the year, they read books and magazines devoted to hunting. Many go to one or more of the annual sportsman's shows that are staged around the region to showcase new products aimed at hunters and to offer demonstrations of techniques by hunters who have achieved notoriety as dog trainers, skilled bow hunters, or expert turkey callers. They watch shows about hunting and fishing on cable television, and many of them have a small library of videos dealing with their favorite type of hunting. Many also belong to local rod and gun clubs where they practice marksmanship, socialize with kindred spirits, and put on game dinners and summertime, family-oriented cookouts.

Among the avid are a few who take their hunting to another level of preoccupation. These are the men and women for whom the challenge inherent in hunting takes on a competitive edge. Unlike George Dunnerston, who didn't keep score, these hunters do keep score, literally and figuratively. They strive for perfection. They have fun, and time in the woods or at the shooting range is recreation, as it is for other hunters, but Charlie Rock, Jon Harcourt, and Cal Jones hunt with an uncommon intensity and focus. They are competitors who prepare for the hunt in much the way an athlete would train for a game. Jon and Cal compete with themselves— they want to improve over their past performance. There's always the prospect of a bigger deer, and there is always the hope of making a more precise shot. Charlie competes with himself, too, but he also competes on the archery range.

Charlie Rock is the most consistently successful deer hunter I interviewed. Since he got serious about deer hunting about fifteen years ago, he has killed at least one deer per season, and he frequently brings home three or more (he hunts in New York and New Hampshire as well as in Massachusetts so he can legally kill at least two in each state). What makes this all the more remarkable is that he hunts exclusively with bow and arrow. Charlie has a small archery shop above his garage. "The little bit I sell basically pays for my equipment and trips to tournaments," he explained as he gave me a quick tour of the benches where he makes custom arrows and tunes up bows. The walls of his shop are filled with medals, ribbons, and trophies he has won in archery contests all across the country (including several national titles in various classes designated by distance and type of bow used). He tries to practice with his bow at least an hour a day, "more when there's an important meet coming up." Charlie looks every bit an athlete. Though he didn't talk about exercise, it was plainly apparent that he works out and runs in addition to practicing with his bow. No one can compete at Charlie's level without paying serious attention to conditioning.

Charlie and his brother began hunting with their father when they were in high school, though then he hunted with a shotgun. He remembers enjoying getting out, but hunting competed with other teenage interests— sports, girls, and motorcycles. After graduating from high school, Charlie tried out for and was invited to join a motorcycle racing team, and he spent the next ten years racing, in the process becoming a highly regarded mechanic. He makes his living now as the service manager for a large New England motorcycle dealership. In the garage beneath his archery shop, he has his own machine shop where he does custom work on bikes shipped to him from all over the world. Parked in the drive was his own bike, a dazzling spotless chrome and fire red machine that looked, standing still, as if it was going a hundred miles an hour. He and a young man whom he was training were at work on a bike when I drove up. After giving the assistant some helpful hints and showing him the tools he would need to do the repair, Charlie took me upstairs for a tour of the archery shop. As I surveyed the trophies and ribbons, he explained how he got into archery.

Charlie had a bad accident on the track that required a lengthy recuperation. Early in his recuperation he stumbled upon a story featuring a top archer and since his arms had been unhurt, it struck him that he could shoot a bow even with his limited mobility. "Hell, I had nothing better to do, and it gave me a goal," he explained. As his body healed and his marksmanship improved, he began to think about hunting. Over the course of his racing, he had rarely had the time to return to Massachusetts to hunt with his father and brother. But, with his racing career behind him, hunting became possible again, and his rapid progress with the bow made hunting suddenly far more compelling than it had been when he was young. Charlie recalled for me the first time he went into the woods with his bow. By then he was fully recovered from the accident and had begun to win local archery matches. He was confident of his skills but, he said knowingly, "there's a whole lot of difference between hitting a bullseye from fifty or seventy-five yards in an open field and hitting a vital spot on a deer in the woods at fifteen or twenty yards."* He remembers passing up shots he now knows he could make because he was not confident that he could place a killing shot. But patience paid off, and a deer finally presented him with a shot he knew he could make. That was the end of shotgun hunting for Charlie.

Several years after returning to Massachusetts, he and his wife moved to

---

*For readers who have never hunted deer in New England, the difference may not be obvious. The forests of New England are more like tangles than glades. It is uncommon to have unimpeded sight lines beyond forty or fifty yards. Add to this the fact that even the slenderest of twigs can deflect an arrow's course, and one begins to see the challenge involved. As if this weren't enough, the optic nerve plays tricks: seeing isn't always believing. When a person is focused on an object, obstacles between the observer and the object tend to be ignored. This is one of the reasons hunters sometimes make unsafe shots: despite all the good intentions in the world, hunters can be so fastened on an animal that they literally do not see anything else. It takes practice and willed discipline to override this perceptual glitch. I discuss this further in the next two chapters when I consider hunter restraint and the issues involving hunter safety.

their present home, located in one of the most heavily wooded sections of central Massachusetts. He literally has prime deer habitat just beyond the edge of his lawn, which extends farther back than he'd like (it's a lot to mow) but it accommodates his archery range. We joked about how if he let his lawn revert, he could shoot deer from his deck. "If I have another accident," he said with a broad smile.

Charlie does not kill deer in order to stack them up. He is not trying to see how many deer he can kill in a season. He hunts deer because that is, so far as he can tell, the ultimate challenge to an archer. Making a good, clean shot that kills a deer promptly is more exacting than hitting targets, even when the targets are at distances that would tax many peoples' eyesight, not to mention their accuracy. Part of the reason involves adrenalin: competitive shooters (and this is, so far as I can tell, as true of firearms as of archery) have their own rituals, much like a baseball player coming to the plate or a basketball player preparing for a free throw, that calm them and allow them to concentrate on one single objective. All extraneous distractions must be blocked out. In the controlled setting of a tournament, this mental toughness is hard to maintain, but like physical conditioning, hours of practice make it possible. Hunting, by contrast, is anything but a controlled setting. The variables are manifold and constantly shifting, and there is always the element of surprise when a deer comes into view. Your heart races, your hands shake with the effort to stay in control as you wait and wait and wait for the moment when you can send your arrow (or shotgun slug, for that matter) on a clear path toward the heart or lungs. The combination of waiting and the prospect of taking an animal's life produces an almost exquisite tension that is not for the faint of heart, and according to Charlie as well as the other, less adept archers I interviewed, is as demanding a discipline as there is.

Charlie hunts deer with the frame of mind of a professional athlete. He is constantly seeking ways to improve his skill and to raise the bar of his accomplishment. He is not trying to be macho, he is trying to be the very best he can be. Trophies and ribbons are one measure of his accomplishment, and the thrill of competition, nurtured first in school sports and later in motorcycle racing, is still compelling. But it is deer hunting that is the most demanding. "Deer are tougher competitors than anyone I face in a tournament," Charlie told me, his voice conveying deep respect for his fellow contestants, both two- and four-legged.

Cal Jones is a carpenter who lives in the Berkshires. He is single and in his mid-thirties but has plans to marry a woman he's been seeing for the past several years, though he's a bit worried that marriage will make it harder for him to hunt as much as he likes. As a single man, he can work when he wants to and refuse work when it suits him. His goals have been to work just enough to support his modest standard of living and to save a little for the proverbial rainy day. Thus far, this arrangement has meant that he can devote a lot of time to preparing for the hunt and to hunting. Eating venison helps reduce his grocery bill, which in turn might, in effect, buy him a couple of additional days without work. In this sense he depends on game for food and in some ways resembles a subsistence hunter, but Cal's near-poverty is chosen: he is voluntarily "poor." To be sure, he acknowledged he has sometimes cut things uncomfortably close to the bone and has occasionally had second thoughts about the lifestyle he has elected. He had an especially hard time a couple of years prior to our interview when he went the whole winter without work and had a slow spring on top of it. "Things were really tight that year," he recalled. He acknowledged that he probably can't go on in this fashion for too many more years, but for now, he remains willing to live on the edge.

Cal consumes deer, but metaphorically, it is equally true to say that he is consumed by deer. He is constantly scouting for places that he thinks are likely to hold a really large buck. In the winter he looks for tracks in the snow and follows those made by the larger animals. In the spring, he makes a point of looking for shed antlers, taking note of the size and symmetry of the antlers. During the summer, Cal's obsession has to take a back seat to work, but he consoles himself—"the woods are too thick to see much and the bugs are even thicker." In the fall, he begins to shift into high gear. There is work to finish up, and while the days are still long, Cal spends his evenings and weekends scouting. He hunts birds during the fall season, but in fact he is really not hunting birds so much as he is looking for deer sign.

Cal mixes a subsistence with a trophy orientation. He hunts in New York State and Connecticut "for the freezer," though the nonresident fees in New York rose steeply the year before I met Cal, and he will probably stop hunting in New York.* His goal is to get two or three deer out-of-state

---

*Both states have much longer deer seasons than Massachusetts. More importantly from Cal's point of view, both states have large deer herds and, in the case of Connecticut, very generous bag limits. Because the Nutmeg State is overrun with deer, it is possible, with all the permits and tags, to kill as many as seven deer per season.

before the Massachusetts season so that his hunting efforts in Massachusetts can be devoted to finding a trophy buck rather than to filling the freezer. Though he depends on venison, he is reluctant to kill does—"kill a doe and you've really killed two or three deer, the doe and the fawns she'd have in the spring." Only in Connecticut does he relax his protective stance toward does because deer are far too numerous there for their own good. Hunting out-of-state tunes him up for the real challenge and the kind of hunting he most prefers: hunting alone and matching wits with the largest bucks he can find.

Many nonhunters assume that men like Cal, who are passionate about getting big bucks, are out to prove their (beleaguered) masculinity. But talking with Cal revealed no desire to dominate. Getting a big buck wasn't a "power trip" for him. Success in this endeavor confirms a combination of woodcraft, marksmanship, endurance, and intimate knowledge of deer behavior and deer ecology. A big male deer is simply the most challenging. Large bucks, most everyone agrees, are the most wary and the most canny of any game animal in North America. They lose some of their wariness during the mating season, but the rut is over by the time the Massachusetts deer season opens. Large deer then become reclusive, resting on their laurels, as it were, and begin regaining the weight they lost during mating. Many good hunters will go for several seasons without ever seeing a big buck; average hunters, those who do little scouting in the preseason and who only get out for a day or two, are lucky if they see one big buck in their lifetime.

If Cal were a mountain climber, he'd yearn for a chance to climb Mt. Everest. Were he a bowler, he'd practice and practice in order to roll a 300. Consistently getting big bucks is the equivalent of such accomplishments. When I interviewed Cal, he was still basking in the satisfaction of having accomplished a really remarkable feat: the previous fall, he managed to bag two bucks, each of which weighed over two hundred pounds, field dressed, and had nearly identical and impressive racks. He shot them on successive days in very nearly the same spot, which suggests they might have been twins. Despite the fact that money is tight, Cal decided to have the two deer mounted. At his suggestion, we drove to the taxidermist's where the trophies were being held while Cal got the money together to make final payment on the work. As we drove, he recounted the details of that hunt, including how he had stumbled upon the place where he killed the deer during his preseason scouting and instantly recognized that this would be a promising spot come the opening of deer season. It is very rare to find an area in which there is clear sign of two large bucks sharing turf. But the sign was unmistakable to Cal, and when all of his scouting paid off beyond his wildest dreams, he was overwhelmed. "It's like winning the Superbowl or something," he explained.

Jon Harcourt has killed a few large bucks in his day but is not consumed by big bucks in the way or to the degree that Cal is. The challenge Jon sets for himself is to be able to get a deer after the season has been in progress for a couple of days. By then, the deers' patterns have been thoroughly scrambled. They avoid their usual haunts, alter their feeding times, and generally just try to make themselves scarce. That's when hunting skills are put to the test. Many hunters grow attached to one, or at most a handful of favorite spots to which they return year after year. I have already indicated how this sort of attachment to the land can be richly imbued with memories and the reassurance of connectedness and continuity. Though emotionally satisfying, this pattern is not necessarily the best hunting strategy. The land changes, and what was prime habitat ten years ago may have passed its prime. Game populations fluctuate from one area to another, and hunting pressures vary. All of this means that hunters who are flexible and adaptive, who know how to "read" cover, and who know a great deal about deer (or grouse or turkey) are likely to see more game than the hunter who returns each year to the same few spots. Jon is adaptive and prides himself on being able to find deer under conditions that would send most hunters home empty-handed.

Jon and his partner live in a rural town in west central Massachusetts. Their house overlooks a small stream that boasts a healthy population of trout. The house started out way back as someone's camp. Jon has put a lot of work and no small amount of cash into making the place comfortable. The décor is unmistakably that of a hunter—a couple of deer mounts and antique muskets on the walls. His partner does not hunt but clearly admires Jon's skill and appreciates the centrality that deer hunting has in Jon's life. Jon grew up hunting, and even before he was old enough to hunt, he was in the woods as often as he could manage. About fifteen years ago, when he was in his mid-thirties, he began concentrating exclusively on deer. He hunts alone and will leave an area, even if it is very promising, if other hunters are in the vicinity. Safety is a concern but not the deciding factor—Jon wants to concentrate totally on the hunt and strongly dislikes being distracted by other hunters. "I just like being one on one with deer," he told me as he described deer as the most beautiful and smartest animal he could think of.

A tall, wiry man in his late forties or early fifties, Jon looks every bit the person who loves battling through the laurel "hells" that cover the hills of west central Massachusetts. How deer manage to move through these tangles is anybody's guess, but since that's where the deer often hole up once the shooting begins, that's where you will find Jon. Jon is not seeking the biggest buck, though he sometimes adds to the challenge by deciding not to shoot at anything less than, say, a six-pointer (a deer whose antlers have a total of at least six tines); but unlike Cal, who almost never applies for a doe permit, Jon regularly applies for one.* It is not that Jon is dying to kill a deer. Instead, he understands and accepts the fact that the herd needs to be kept within the bounds set by the availability of habitat and the willingness of humans to tolerate deer. That means that killing does in regulated numbers enhances the long-term prospects of the deer herd.

Jon sustained a work-related injury several years back that ended his bow hunting, though he holds out hope that he will be able once again to hunt with a bow. The attraction of bow hunting, as many others noted, was the quietness. One can see more deer, even if they never get within range, and one can see a lot more wildlife. The next best thing is hunting after the noise and commotion of opening day is past. The hunters thin out, and the deer spread out; if it is not exactly quiet, things calm down. And if one heads for the really thick stuff, he or she is not likely to encounter many other hunters—and those encountered are likely to be serious and thus knowledgeable and safety-conscious.

Jon prefers to hunt by himself, though he does go out with an older friend who has diabetes, which precludes him from hunting alone. When they go together, Jon and he equip themselves with radios, just in case. In general, though, Jon spurns anything that gives him unsporting advantage over the deer. He does not do a lot of preseason scouting because, as I've

---

*In most states, the taking of does is regulated according to the size and health of the deer herd. Since deer are polygamous, and since it is virtually impossible to kill all the males in a given area, even if mortality of bucks is quite high, as long as there are plenty of does there will be robust reproduction. Doe permits are issued to reduce or stabilize herd size. In some states like Connecticut, where the deer population has reached levels that pose hazard to humans as well as to the deer themselves, hunters are encouraged to kill does. In some areas of Massachusetts, notably Martha's Vineyard, the same is true. Still, many hunters resist killing does. This reflects the combination of a bias toward bucks because they are more difficult to find and are thus a sterner test of a hunter's ability and the collective memory, passed down from generation to generation, of a time, fifty years ago, when deer were really scarce. Old habits are slow to change.

already noted, once the shooting begins, the preseason patterns dissolve. He also does not do any shooting before the season other than firing a few rounds just to make sure the sights have not been knocked out of alignment. In these respects, his preparations for hunting are minimal compared to the hundreds of hours Charlie spends practicing with the bow, or the similar amount of time Cal invests in scouting for deer. But each in his own way comes at hunting with a studied determination to make the challenge as demanding as it can be. No shortcuts, no "cheap shots," nothing to take anything away from the dignity and nobility of the animal they think is the most marvelous creature on four legs. They approach hunting with a seriousness that most people would associate with religious observance or some solemn responsibility.

Taking Fun Seriously

In practice, each hunter has fun hunting, enjoys recreational benefits, and experiences at least some measure of the challenge of sport. If it were work, or if it were driven by harsh necessity, I am sure that the men and women with whom I spoke would have given me very different accounts of their activities. An experience that Jon Harcourt related to me is revealing in this context. Years ago, he worked for an orchard grower.

> He asked me if I would shoot deer for him. I said, "Sure." He got the permit for taking deer out of season. Oh man, I was excited. I thought this is going to be a blast. I got up on his barn with a high-powered rifle, dusk came and, boom, out came the deer and, boom, down they went. And then I told him "that's it, I ain't doing that no more." "What do you mean," he asked? I explained that I don't like to shoot deer, I like to hunt them. And that was the last shot I ever fired in that orchard, even during the season. And he was mad because he really wanted me to do the killing.

When killing an animal becomes work, when there's no challenge, no fun, it is simply not hunting. What little we know of subsistence hunters suggests that even though the practical stakes were high, excitement and fun lent their hunting much the same sort of psychic rewards as the hunters I interviewed experienced. For example, Richard Nelson (1990), an anthropologist who has lived among the Koyukon people in Alaska for many years, describes in vivid detail his attempt to emulate their hunting practices. There can be no doubt that the hunt was as filled with drama and adventure for the Koyukon as it is for Russ or Jon or Charlie. It is also clear that the Koyukon look forward to the coming of the hunting season (not the season set by game laws but rather the season tradition has established), not least because it signals a change in activity and, as we would put it, a chance to "get away." Another anthropologist, Hugh Brody (1982), provides us with a detailed account of the hunting and fishing of native peoples in British Columbia. Here, too, it is clear that the hunt represents more than a routinized acquisition of food. Ted Kerasote (1993), noted outdoor writer, has also shown that the subsistence hunters of Newfoundland get more than protein from their hunting. The risks, adversity, and discomforts they routinely confront confirm their worthiness as breadwinners. At the same time, however, the hunt dramatically breaks up the boring routine of hanging around the village with "nothing to do." Of course, there are many things to do, and the hunters are quite busy with them. But it is the busyness of small and unexciting things—laying in firewood, repairing equipment, and household maintenance. If there is any remaining doubt about this, then surely the tales passed on from one generation to another recounting the exploits of legendary hunters, layered through with myth and embellishment, clearly signify that hunting is never just about getting bellies full.*

In precisely this way, the distinction between modern sport hunting and traditional subsistence hunting is far less stark than it would seem at first glance. Despite the sweeping differences between our lives and the lives of our hunting-gathering forebears, the motives of hunters have changed remarkably little over time. Even the idea that hunting was necessary no longer seems compelling. No doubt there were some peoples who were more heavily dependent upon the hunt than others (the Plains Indians' close link to the bison is a case in point), but anthropologists and historians now think that gathering activities and horticulture (including the rather undramatic snaring of small mammals and birds) figured far more crucially in the diet of hunter-gatherers than had long been assumed. For

---

*Not all subsistence hunter-gatherers hunted in the sense in which I am using the term. For example, some Plains Indians stampeded bison over cliffs and then, when the dust settled, they dispatched the animals that were not killed outright and proceeded to butcher all they could handle, leaving the rest to the wolves, vultures, and others lower in the food chain. This process strikes me as more like work than like hunting. It would be nice to know how they compared that mode of obtaining sustenance with the ways they acquired other game.

a very long time, hunting has been inspired at least as much by symbolic needs as by nutritional needs.

It is easy to overlook this past and to conclude that nutrition (necessity) gives license to kill, but the pursuit of fun, recreation, and sport do not. In light of this spurious conclusion, it is not surprising that large numbers of people have come to regard hunters as brutes. After all, who but a brute would take pleasure and find relaxation in pursuing splendid creatures? How can hunters kill "for the fun of it"? I hope that it is now apparent that the "fun" hunters enjoy is far from trivial fun—hunting is not like a day at the beach or a game of golf, though there are elements of hunting that are shared with these other sorts of diversions. Hunters do have fun, but they are not hunting for the hell of it. Hunting is serious fun. Beneath the smiles and laughter and backslapping, there is a solemnity that is irreducible. That's because hunting is about killing lovely animals. How can any decent person, especially if he or she is not starving, do that?

V

KILLING GAME

Killing an animal is an emotionally complicated and conflicted act. Everyone who has even casually read the literature of hunting has encountered descriptions of how the successful hunter, at the moment of the kill, is bathed in a flood of conflicting emotions—triumphal elation alternates with feelings of remorse, and both are mixed with awe and humility. Indeed, among the ironies that confound critics of hunting, the fact that hunters, ancients as well as contemporaries, profess a profound admiration for, even love of, their quarry is the most difficult to fathom. The attributes of cunning, majesty, and sheer beauty that hunters attribute to game animals surely make it harder to explain why such magnificent creatures should be killed and raise the possibility that the remorse attending the kill is the precursor of a bad conscience. Historically, the many and varied rituals that framed the hunt worked to absolve the hunter of this guilt. They were an expiation, a socially agreed upon way of making the killing acceptable. By ceremonially marking the hunt as a special, bounded act, hunters were given more than the strength and courage needed to hunt: they were given permission to kill.

Rituals did not simply assuage the hunter's guilt, however. They also assuaged another sort of anxiety: the anxiety that the hunter, awash in blood, would be transformed into a dangerous person, a person no longer fit for

125

life in the village. The rituals, this is to say, made only certain kinds of killing acceptable. Even as they declared a profound connection between the human and the nonhuman worlds, the rituals of the hunt drew boundaries beyond which killing is unacceptable. Even in cultures in which hunting was critical to survival, the rituals marking the transition from villager to hunter and back to villager reveal these two related anxieties. Part of the reason hunting is now attracting mounting criticism is that we have long since abandoned the public or communal rituals that reassured both hunters and nonhunters. To be sure, in rural areas, the opening of deer or elk season remains something of a public event. Churches, local merchants, and the schools take note, accommodate hunters with send-off breakfasts and otherwise wish them well (Miller 1992). But, for the most part, as in so many other respects, the bank of our common rituals, our "ritual commons," is very nearly broke. What remains are pallid, faint shadows of the fasts and feasts and supplications that marked the beginnings and endings of the hunt. This has meant that hunters and nonhunters have long since ceased sharing a common understanding of the hunt and of one another. More importantly, hunters now have no way of reassuring nonhunters of their membership in a common moral universe.

This loss of public ritual has affected nonhunters far more than it has affected hunters, not least because hunters have replaced traditional rituals with an ideology that reassures *them* of the moral acceptability of what they do. The problem is that hunters have no way of reassuring the nonhunting public of their probity. The public's anxieties are not allayed, and when these anxieties get combined with increasing solicitude for animals, public support for hunting necessarily declines. I shall examine this dynamic in greater detail later, after I have explored the ways contemporary hunters frame the hunt in order to assuage their guilt and erect boundaries such that the passions of the hunt do not spill over into daily life.

Framing the Hunt

There are three distinct but overlapping features of the way hunters contextualize hunting. Some emphasize one feature more than the others, but every hunter I interviewed mentioned each feature, almost as though it were a catechism. The three components that constitute what might best be thought of as the "ideology of contemporary hunting" are: (1) when I hunt, I leave civilization and enter the realm of nature, where different

126

rules apply; (2) I endeavor to make a clean kill; and (3) by practicing self-restraint (the "fair chase"), I achieve a sustainable relationship with nature. Let us examine each of these claims in turn.

Entering Nature. Hunters routinely draw a sharp distinction between society and "the woods." Indeed, it is precisely the difference between these two realms that most hunters find alluring and refreshing. As I have shown, hunters value "getting away," by which they clearly mean getting away from the constraints, constrictions, and conventions of daily life. In nature, they feel free, constrained only by the limits imposed by the environment and by their physical capacities. In society, hunters are husbands, wives, sons, and daughters; they are workers (or students hoping some day to become workers); they have bills to pay and appearances to keep up. Their lives are scheduled, usually to accommodate coordination with many other people with whom they live and work.

By contrast, when hunters are in the woods, they are on their own. Even the hunters who hunt with a consistent group of friends or relatives note that once they leave the camp or their vehicle, each becomes his or her own moral agent. It is up to each, and each alone, to decide to shoot, or change positions, or make any of the countless small decisions that cumulatively add up to having hunted. There can be group strategies, but there is no close-order drill. Nature is too unpredictable, too varied, to yield to tight coordination. For example, bird dogs are bred and trained to systematically work through a field or wood lot, lest the whiff of a grouse or pheasant escape their flared nostrils. In training exercises, the dog and handler can resemble a close-order drill. In hunting situations, though, it is the foolish hunter who insists that his or her dog keep to the measured sweeps back and forth that mark the training process. Scent is mysterious, the birds are elusive, and so the good hunter follows the dog—and the dog follows its honed instincts. Method is punctuated by randomness; spontaneity, not predictability, is rewarded.

Entering nature means entering a world in which rules very different from the rules of society are at work. Most dramatically, the hunter enters nature as one more predator operating in a realm where everything is food for someone or some thing. Peter Carbone underscored this point dramatically. He observed that going out prepared to kill an animal changes your relationship to wildlife and to nature, whether or not you actually kill anything and even if you do not discharge your weapon. Being prepared to kill an animal means the hunter becomes a participant in nature, not simply a spectator.

In nature, the rules are cast in stark Darwinian terms: survival of the fittest is the way things work. Nature plays no favorites. Hunters are acutely aware of the fact that they are competing with other predators.* In this context, killing and death are normal. In effect, hunters say to themselves, "Were I not to pull the trigger, the bird's or deer's life expectancy would not be appreciably altered." Longevity and rates of mortality are framed by habitat, climate, disease, and nonhuman predation. Hunters believe that they account for less mortality in most hunted species than the combined effects of the other sources of "natural" mortality, and were hunting to cease, the creatures thus spared would not make it to the next breeding season. Moreover, unhunted populations, even when predators are present, do not stabilize at some golden mean. The moose population on Isle Royale, for example, probably the most thoroughly studied in the world, fluctuates dramatically, as do their predators, the wolves. But the fluctuations are by no means synchronized—rather, they are more nearly random (Botkin 1990).

Few hunters read scientific studies. But the "hook and bullet" magazines they do read regularly report on research findings that summarize the factors affecting mortality, so-called game cycles, and the role that hunting plays in the population dynamics of various game species. In general, these studies rank hunting very low on the list of factors causing game animal mortality and confirm the view that nature is, indeed, a system ruled by predation. Moreover, hunters see abundant evidence of violent death—piles of feathers that mark a red-tailed hawk's meal, the scat of a fox or coyote filled with rabbit fur, and the bloodied bits of hide and bones scattered over the scene where coyotes feasted on a deer. Nature's luxuriant fecundity is matched, must be matched, by an equal extravagance of death. Hunters accept this and think of themselves as just one more element in this natural cycle of life and death.

Hunters think of themselves as joining the ranks of predators, but they also acknowledge that, unlike a hawk or a coyote, indeed, unlike their dis-

*Some of the hunters I interviewed said that they had shot, or would shoot if they had the chance, coyotes in order to relieve pressure on deer and turkeys. More interesting, though, many more hunters said they didn't mind sharing the woods with other predators and had no interest in hunting them.

tant ancestors who were as dependent upon prey animals as were hawks and coyotes, they are not compelled by necessity to hunt. No one I interviewed would have starved had he or she not hunted, though for a small handful, the fish and game they brought home made making ends meet somewhat easier. (Had I interviewed hunters in northern New England or in the upper Midwest, rural South, or the Mountain states, I would no doubt have found more people living off the land.) Hunting was a conscious choice borne, not of necessity, but of a desire for the experience of being an active and engaged participant in nature. The fact that this is a choice makes it necessary for hunters to do more than don the identity of predator. Killing may be normal, but *how one kills* is also crucial.

The Clean Kill. Killing in the hunt is quite unlike killing a domestic animal. No one defines the latter as "fun" or "exciting." It is precisely the excitement and the passion of the hunt that is troubling, for both hunter and nonhunter. Just as gynecologists and their patients define the pelvic exam in completely nonerotic terms, thus permitting intimate contact that in any other context would be utterly wrong (Henslin and Biggs 1995), so hunters define their killing in ways that remove any hint of bloodlust or primal violent impulse. There are two key elements of this symbolic redefinition of killing: the clean kill and the self-restraint of the fair chase.

Hunters again and again emphasize the lengths to which they go to honor the dignity of the animal they are pursuing. Central to this respect of game is the commitment to careful shot selection in hopes of effecting a quick, clean kill. To underscore the point, several hunters recounted instances in which their judgment had been bad, and the result was a lost cripple. They made clear in no uncertain terms how deeply they regretted their error. Carl Osgood recalled one of his earliest hunting trips accompanied by his father. Eager to "get my deer," he rushed his shot and hit the deer in the gut—a bad shot in every respect: the animal will be able to cover considerable distance before it slowly and painfully succumbs to its wound; this significantly reduces the likelihood that the deer will be found; and, if it is found, much of the meat will have been ruined by being contaminated by the contents of the stomach and intestines. Carl recounted how his father upbraided him by describing in gory detail what the deer was going through, and to drive the point home, he insisted that he and Carl find the deer. After an exhausting day of tracking the wounded animal, with light rapidly failing and several miles from camp, they finally found the severely weakened but still living deer. Before dispatching the deer, Carl's father once more asked him to imagine the agony he had imposed upon this magnificent creature. As he recounted this sorry tale, Carl hung his head, no doubt the posture he assumed when his father was admonishing him. The years had not dulled the shame and remorse he felt that day long ago. Carl ruefully observed "that was one deer I was not proud of," the understatement making the point all the more forcefully.

All but the most casual of hunters (those who go out once or twice, more to get away and be with friends than to bring game home) spend some time in the off-season scouting for game, keeping themselves and, for the bird hunters among them, their dogs in reasonable physical condition (only one of the nearly forty men and women I interviewed was notably overweight; most were in fact quite trim), and honing their skills with shotgun or bow at the shooting range. These activities were fun in their own right, but the larger meaning they held was that these were ways to ensure that when presented with the opportunity to kill an animal, the hunters could effect the kill as cleanly as possible. To do otherwise would be irresponsible and demeaning of the animal.

No one talked with relish about the killing. By contrast, hunters were intense when they spoke of the times when they heard or saw a deer or a turkey approaching: as their adrenalin began flowing, the animal turned away or stopped just at the point where a tree or brush obscured it enough so that a sure killing shot could not be made. Peter Hanks's account of his first turkey hunt captured this nicely. It is worth recalling his words. He was recounting how the turkey he had been calling came in sight but was still too far away for a sure kill shot. Peter waited breathlessly, only to see the turkey look in his direction and veer off. He explained: "I'm not out there to make an animal suffer. If I can make a clean kill, then that's exactly what I'm going to do, and I practice so that I can do that. After all, I don't need to kill that bird." He went on to liken his waiting for a good kill shot to the way predators stalk their prey. Predators live on the fine line of having to expend more energy in capturing a meal than the meal will yield in calories. Peter had read that wolves, for example, will break off a chase if the outcome does not look promising. For him, calories weren't the bottom line; conscience was. By not taking a low-probability shot, Peter was affirming the worthiness of the bird, the sight of which was gratification enough. As importantly, he was making clear that he was a hunter, not a killer.

In one way or another, hunters said that they preferred to hunt well and come home empty-handed than to come home with game that had not been killed swiftly. Every hunter with whom I spoke dreaded wounding an animal and not recovering it. Better to miss; even better not to have shot at all. A number of hunters contrasted their dedication to achieving a clean, quick kill to the often violent and protracted deaths animals are subjected to in the wild. Wolves begin tearing an elk apart well before the animal loses consciousness. Rabbits and squirrels are carried away in a hawk's talons still very much alive. Compared to this, a well-placed arrow or pattern of shotgun pellets brings death quickly. If animals will die in any event, hunters reason, it is surely the case that death at the hands of humans is quicker and more humane than death by fangs, claws, or talons. Hunting, from the perspective of hunters, imposes no gratuitous suffering on wildlife and may in fact reduce net suffering. This is certainly arguably the case with species like white-tailed deer, who, absent hunting, will reproduce until their numbers exceed the carrying capacity of their habitat, which ushers in a cycle of starvation and disease that reduces the population far below the levels that could be sustained were hunting allowed. Hunters are predators, but they bring to their predation weapons and practices that, in their mind, minimize animal suffering. The goal is a shot that delivers a quick kill. Meeting this goal requires self-restraint.

Restraint and Sustainability. Endeavoring to achieve a quick, clean kill means that hunters have to pass up shots they judge to be too difficult or too likely to cripple but not kill. This is a matter of judgment, and there are plenty of tough calls involved. First, skills vary. For a champion archer like Charlie Rock, a shot that Peter Carbone would not dream of taking might well be a kill shot. Hunters can come to know what is a good shot for them—their limitations—only by practicing. Then they must have the good sense to recognize those limitations when the time comes to decide to release the arrow or squeeze the trigger. It's one thing to shoot at a target at the shooting range. It's quite another thing to shoot at a deer or a duck or a pheasant. Some people snap with the excitement—hence, "buck fever." I have no way of knowing how many hunters lose control and blaze away in a paroxysm of excitement. None of the hunters I interviewed mentioned that happening to them. But it certainly does happen. Such loss of self-control is nothing to be proud of, which means that hunters are not likely to freely admit that it has happened to them. The point is not that hunters may be concealing something unflattering about themselves (I will address this shortly from a somewhat different angle). Rather, the point is that the disavowal of buck fever underscores the importance of the ideal of self-restraint. Hunters do not see themselves as impulsive people who lose control and spray the woods with lead.

The way hunters support this claim is by noting the frequency with which they pass up shots and how reluctant they are to press their luck with a low-probability shot. Jack Wysocki mentioned his reluctance to take a shot at an elk that was, as he put it, "out of my comfort zone," by which he meant that the elk was too far away for Jack to be confident of being able to place his shot well. Paul Julian passed up a "decent shot" in hopes that the deer would come a bit closer and fully into the clear. Instead, the deer vanished, as if into thin air. Russ Farina, the epitome of a recreational hunter, spoke of how he gets satisfaction out of seeing game and being able to resist the impulse to shoot until everything is just so. Mark West and Dick Board, the bird hunters, contributed another variation on this theme. Both hunt with dogs bred and trained to point birds—to freeze when their nose tells them a bird is nearby. The hunter then moves ahead of the motionless dog to flush the bird. Sometimes even the best of dogs will "bump" a bird, usually by crowding it, and forcing it to flush. Mark and Dick both said that they will not shoot at the bird when this happens. They only shoot at birds their dogs have staunchly pointed. In his retirement, Dick has begun hunting with friends who have not hunted over pointing dogs, and he noted that he had to remind them that they should not shoot at birds that his dog has not properly handled.

Taking care to select shots is a key element in the ideal of the fair chase: by minimizing the likelihood of crippling, and thus minimizing the animal's suffering, the hunter is honoring the quarry. The restraint involved in passing up shots not only demonstrates regard for the individual animal being hunted, it is also a way of underscoring the claim that sport hunting is based on maintaining a sustainable relationship with game populations. If hunters simply blazed away at any and all targets, game populations would undoubtedly suffer along with the suffering of individual animals. The stories hunters shared of how they have passed up shots either because something was not quite right or because they simply got carried away with the pleasure of watching an animal in its natural state demonstrated that they are not trigger-happy goofs. Among the hunters

I interviewed, there was a general agreement that so-called game hogs, people who are hell-bent on bringing home as much game as they can, are deplorable. Waiting for the clean kill shot is about more than the desire to be a discriminate and humane predator. This sort of self-restraint is also about not being greedy, which is to say, it is about solicitude for the animal and concern for the health of the herd or the reproductive potential that will ensure robust numbers into the future. Even when game laws do not require it, hunters are reluctant to shoot does or hen pheasants. Most bird hunters, at least those who hunt with a dog, will refrain from shooting a bird on the ground or sitting on a tree limb. "I only shoot birds that my dog has worked," is a common refrain among bird hunters. Hunters who specialize in woodcock quite commonly say that because the population is declining, they shoot fewer birds than the law allows. Indeed, in recent years, publications aimed at bird hunters have begun to carry stories of hunters who have decided to quit shooting woodcock, even though studies show that habitat loss, not hunting pressure, is the decisive factor driving the population decline.

Hunters are proud of their collective record—game species that, fifty years ago, were scarce or nonexistent locally are now abundant. And those species, like the woodcock, whose numbers are declining, are being rallied around. Of course this is born of self-interest—but it is certainly a near-perfect example of enlightened self-interest. Were this sort of self-interest governing cod fishing or the taking of tuna and swordfish, those species would now be thriving rather than threatened. Hunters do grumble about this or that regulation, but they do not bristle at the principle that the taking of game should be regulated. They understand and fully accept the fact that regulation, if not every specific rule, is what ensures robust populations of wildlife into the future.

Framing the hunt in these ways allows hunters to kill animals with a clear conscience. They do not go into the woods to inflict gratuitous pain and suffering any more than they go into the woods thinking that animals are dispensable. They hold wildlife in high regard, and they seek to do all they can to ensure that wildlife prospers. From their point of view, predation is crucial to maintaining healthy populations, and as a result, there is no contradiction between killing and preserving, so long as the killing is regulated both by game laws designed according to the evolving principles of game management as well as by the personal codes of self-restraint and fair chase. Still, there is the killing.

133

"Killing's not the fun part."—Karen DeFazio

Only one of the hunters with whom I spoke talked about killing an animal with little or no affect. Tony Vitale, a retired butcher, had been around dead animals his entire adult life. He had little sympathy for and, so far as I could tell, scant sentimentalization of the animals we regard as food, including the game animals he hunted. In this, Tony was the exception. Like every other hunter, he sought a clean kill—a shot that would bring the sort of instant death that is achieved in the controlled setting of a slaughterhouse. But unlike the other hunters, his desire for a quick, clean kill had less to do with solicitude for the animal or a general desire to avoid unnecessary suffering than with concern for the quality of the meat. An animal that takes a long time dying, he explained to me, releases so much adrenalin and other hormones into its blood system that its flesh is unpalatable. (This may in fact be one of the sources of the taste so many attribute to game and that turns off many people to game.) Tony, I hasten to add, was anything but hard-bitten. Though his language was coarse, he had a well-thought-out position: one only shoots what one can and will eat and one must take pains to make sure that the meat will be delectable. This is how Tony shows his respect for the animal. Killing is a necessary fact of life. It should be done expertly: swift and sure, a sacramental as well as a provisioning act.

Because of his skill at butchering, friends often ask Tony to butcher their deer, in exchange for which he is offered a choice cut. He can tell from the condition of the carcass whether his own standards have been met; he refuses to keep meat from those deer that did not die quickly or were not properly field dressed. Tony has a low opinion of hunters who botch things, especially those who don't even seem to know that they have messed up. He reserves special disdain for the hunters who use buckshot instead of slugs. Buckshot typically causes multiple wounds and thus ruins much more meat than a single shotgun slug. Moreover, the hunter has little control over where the buckshot goes once the pellets leave the barrel. By contrast, shotgun slugs can be almost as accurate as rifle bullets at short distances, which means the hunter can be far more selective about the shots he or she takes. The virtues of slugs are crystal clear to Tony: a quick kill and minimal damage to flesh.

Aversion to buckshot was near unanimous among the men and women I interviewed. Some favored an outright ban on buckshot while others simply stated their personal preference for slugs. The reasons they offered included concern for the meat lost, but that was a decidedly secondary consideration. The rejection of buckshot involved safety concerns and the difficulty of ensuring a clean, quick kill. Nearly half of the hunters acknowledged that they had used buckshot in the past, usually when they had just begun hunting and were unsure of their skills or were, like Tommy Worthington, preoccupied with proving themselves worthy to be included in the company of their elders. If that is the goal, buckshot makes sense. Of the younger hunters I interviewed, only one used buckshot. Ian James had taught himself to hunt and then convinced a friend of his father's who had never hunted to take up the sport. Together, the young man (Ian was in his mid-twenties when I interviewed him) and the older man hunt pheasants and deer. Ian allowed as how it was frustrating at first because all they had was the information conveyed in the hunter-education course they took and in the books and magazines they read. As useful as these sources of information and guidance can be, they are no substitute for experience. Buckshot increases the hunter's chances of bagging a deer. Still, Ian and his friend rely on buckshot only as a backup—each uses a semiautomatic shotgun, and they load their guns to first fire a slug, then a load of buckshot, and finally a slug (to administer the coup de grace should it be necessary).

Hunters understand that what they do causes pain and suffering. Hunters are also participants in a culture that increasingly holds pain and suffering unacceptable. While they do not take this to the lengths that animal activists do, hunters are nevertheless bothered by the thought that they have caused an animal to suffer. This sense of responsibility produced some of the most intense comments my questions elicited. People told of how they relentlessly tracked a wounded deer until they caught up with it and dispatched it, or until they were satisfied that the wound was a superficial one from which the deer would recover. In this context we should recall Roland Morreau, the man friends would call upon to help them track a wounded deer. He was understandably as proud of his tracking abilities as he was of his shooting. Everyone who hunts learns that it is wrong not to do all that can be done to track down a wounded animal. The stories I was told included many in which a well-hit deer would bound out of sight and then conceal itself to live out its last moments and be virtually impossible to find, all within a hundred yards or so of where it was shot. Jon Harcourt recalled the time he shot a deer and knew it was a killing shot, even though the deer ran off. He waited for twenty minutes or so before he began searching for the animal.* He followed the blood trail until it stopped and then began to search in expanding concentric circles from the last drop of blood. Darkness began to overtake him, and finally, at last light, he marked the spot where he last saw blood with surveyors' tape; he walked disconsolately out of the woods. He returned the next morning at first light and went directly to the marker. As he surveyed the scene, the reflection of the sun off a shiny object caught his attention. There, not twenty yards from where he had stopped his search the evening before, lay his deer, now quite dead, in a shallow depression in the earth. He had seen the sun reflecting off one of the animal's hooves.

Stories such as this one reveal how hunters come face to face with the consequences of their action. They are forced to witness the incredibly tenacious life force animals possess, and they are also reminded that even a perfectly placed shot does not always result in instant death. Even a mortally wounded deer will struggle to escape, and even a swift death can be accompanied by heart-wrenching death throes. Hunters cannot avoid the violence that is inherent in hunting. By all accounts, this is a very sobering confrontation. The fact that it is mixed with elation seems to add to the solemnity of the moment. The embrace of violent death is troubling to hunters, even though they understand that death itself is simultaneously the end and a renewal, an essential loop in the circle of life. Still, killing is troubling, and this is one of the reasons older hunters report no longer being as interested in killing as they were when they were younger. In my sample, the only exception to this pattern, which has been reported by other researchers as well (Jackson and Norton 1987), was Peter Hanks, who had returned to hunting after his retirement. Bill Crafter, also a retired schoolteacher, is more typical of older hunters. He tags along with his son but acknowledges that he is no longer interested in shooting deer or birds—"I've shot my share; it's for the younger people," was how he put it. He still engages in some target shooting and has no desire to sell or give away his guns. He is simply no longer eager to kill. "Oh, once in a while when the dog (his son's German shorthaired pointer) works a pheasant really well and I'm the only one who has a shot, I'll take the bird. But it's for

---

*This is the accepted practice. If immediately pursued, even a mortally wounded deer will often be able to muster enough energy to make a desperate run. If there is no pursuer, however, the animal will stop fleeing and try to conceal itself. A mortally wounded deer will then simply subside, almost as if it has "gone to sleep." Thus, the wait before beginning the search.

the dog, to honor its good work, not because I want to shoot that bird. I'd be happy just admiring the dog's work."

Albert Cazminsky, a retired maintenance man, still hunts for deer each fall but hasn't shot at one for some time. "I like watching them," he explained, "and I like eating venison, but someone in my family always gets a deer so I get some venison. If I don't get one (a deer) it doesn't bother me. I don't need to shoot a deer." Albert lives on the outskirts of a small city along the Connecticut River; his backyard ends hard against a steep hillside that is owned by the city and managed as watershed land. His large and well-kept yard is dotted with bird feeders guarded by Have-a-Heart traps. ("I used to shoot the squirrels, but now I just trap 'em and take them back up the hill. Good exercise for me and them, I guess.") He and his wife are regularly entertained by the wildlife that venture into their yard from the hillside—turkey, deer, and fox join the woodchucks, squirrels, and rabbits as regular visitors. More recently, coyotes, bear, and even a moose have joined the entourage. Albert, a lifelong resident of the area, reflected back to his youth on his family's farm. "There were lots of pheasants and rabbits, but that was about all. Maybe there were bears in the Berkshires, but there weren't any here and there weren't many deer either. All shot out." He is thrilled by the recovery of so many wildlife species but is upset by the surging population of coyotes. They decimated his flock of domestic geese and ducks, and he did shoot one in his backyard after he lost the pet geese. He worries that there won't be enough hunters to keep wildlife populations in check. "It's getting so there's almost too many bear and deer now. There's going to be trouble."*

The last deer Albert shot was in Vermont. It was bow season, and the day was quite warm. He shot a deer, and as he waited to begin trailing, he began to worry that the heat and the hilly terrain were going to make things risky for a man of his age. As luck would have it, just as he began trailing the deer a young hunter came down from his tree stand. The young hunter had seen Albert's deer go by, clearly losing energy. Albert considered things for a moment and offered the deer to the young man. He accepted

---

*Two days before I interviewed Albert, a bear cub had climbed the fence of a day-care center's play yard, just down the street from Albert's house. Fortunately, the children were inside at the time. They watched spellbound as the frantic mother bear tried to get the cub out of the fenced-in area. Had some children been in the play area, something bad might well have happened.

the offer, took up the trail, and found the deer dead a quarter of a mile away. "Chancing upon him was a godsend," Albert said. He felt honor-bound to carry the hunt to its proper conclusion but knew he risked exhausting himself and a possible heart attack under the circumstances. The young man saved the day.

Rod Majors, the only hunter I interviewed who was far more interested in guns than in hunting, is a strapping man who looks far younger than his sixty-six years. Like Bill Crafter, he goes hunting with his son a couple of times a year to share the time with him, but with no intention of shooting a deer. He takes far more pleasure in shooting at targets, which he does several times a week, year-round. It may strike the uninitiated as strange that someone would love shooting and not like killing, but this combination is common. Many serious shooters are intent on improving their scores. This requires intensity of concentration and, insofar as humanly possible, the elimination of all extraneous factors that might impair one's concentration and accuracy. Hunting is far too unpredictable for such shooters, and the obstructions to a clear sight line, the long waits that invite the mind to wander, and the evasive behavior of the "target" are enough to drive many shooting enthusiasts bonkers.* Rob wasn't bonkers. He simply preferred bull's eyes to a buck's heart or lungs. "I really have no desire to kill something," he said matter-of-factly.

The moral weight of killing makes hunters more, not less, thoughtful. As Jon Harcourt, an accomplished deer hunter, put it, "it's [killing] a necessary evil." He said this without shame or remorse in the course of deploring the "Bambi syndrome." It wasn't the sentimentalization that bothered him. He, along with most hunters, sentimentalize wildlife, too. He was despairing of the way people avoid thinking about death. No hunter can avoid thinking about death, though there are variations among hunters who pursue different kinds of game. Deer hunting draws the highest number of hunters, but many pursue creatures that are less commanding than deer and whose killing may be less morally troubling—putting a squirrel in your sights is not as overpowering as looking down the barrel of a gun

---

*This is one of the reasons why sharpshooters are not nearly as effective as one might think when it comes to reducing the size of a deer herd or killing a problem animal. Consistently hitting targets on a range is far from easy; doing the same in the field is even harder. Even though both require similar skills, being good at one does not smoothly translate into being good at the other.

at a deer. Deer hunters may wrestle with the Bambi syndrome, but no rabbit hunter contends with a "Thumper syndrome." Even though turkey, geese and the upland birds, quail, ruffed grouse, woodcock and pheasants, have their enthralling characteristics, the men and women who hunt them do not generally report the same intense flood of emotional crosscurrents that those who hunt deer routinely describe.

Many of the people who hunt deer also hunt small game, or they hunted small game until deer hunting became a preoccupation, but when we talked about how they felt when they pulled the trigger and saw the animal hit, they invariably talked about deer, not the birds or small mammals they shot. The three people I interviewed who specialized in upland bird hunting, Mark West, Karl Woichek, and Dick Board, did not say so explicitly, but nevertheless, it was clear that one of the reasons they preferred bird hunting to deer hunting was precisely that killing a bird and dressing it out was easier, not only physically easier but, more important for my purposes, easier on one's conscience. Birds just do not have the moral standing of a large game mammal.* It is not that bird hunters are more callous than their deer-hunting counterparts, or indifferent to suffering or to the death of an animal, or that they are any less committed to the goal of careful shot selection and a clean kill. In fact, bird hunters have to deal with suffering and death far more frequently than deer hunters. A highly skilled and knowledgeable deer hunter, someone like the archer, Charlie Rock, or the shotgunner, Cal Jones, might kill as many as four or five deer in a good year. But one deer every couple of years is considered by most to be good, and many deer hunters go for years without killing a deer. (In Massachusetts, about 10 percent of all deer hunters are successful in any given year.) By contrast, an avid and skilled bird hunter who hunts waterfowl or upland

---

*Some would argue that a life is a life, and that the death of a bird should weigh as heavily on the scales of moral reasoning as that of a white-tailed deer. This position has logical consistency in its favor, but logical consistency has almost no bearing on cultural standards. In fact, I would argue that a logically consistent culture would be a perfect nightmare. It is clear that most people, hunters included, do rank species in terms of how compelling they are and how much emotion is invested in their lives and deaths. Running over a squirrel or a raccoon is jarring and unsettling, but it is not nearly as traumatic as hitting a deer—and not just because the deer causes damage to the vehicle and possible injury to the occupants of the car. PETA encounters this problem all the time in their campaigns to save all animals. Most people cannot get exercised about pigeons or doves. That's why PETA and other animal rights groups adopt "poster animals" that are decidedly charismatic—baby seals, wolves, and bears, for example.

birds or both can typically bring home fifty or more birds of various species in an average season. Bird hunters, and small-game hunters generally, have many more opportunities to shoot than do deer hunters.

In the nature of things, quick, clean kills are also harder to accomplish with small game. Most deer are shot when they are standing still or moving very slowly through the woods. By contrast, birds are shot in flight at speeds ranging from twenty miles per hour to a zinging forty to fifty miles per hour. Hunters use shotguns loaded with bird shot of varying size, depending upon the bird being hunted (waterfowl require larger shot because their feathers are harder to penetrate), that begins to disperse in a widening circle as soon as the pellets leave the barrel. This dispersal means that even a well-placed shot will typically hit the bird with only four to six pellets.* This is enough, usually, to drop but not necessarily to kill the bird. Fewer than three or four pellets, unless one breaks the bird's wing or penetrates its skull, usually means the bird will fly off, sometimes with the hunter not even aware that the bird has been hit. Birds can recover from such wounds as long as they are superficial, but if the stomach cavity has been hit, infection will almost certainly kill the bird. There is no way of knowing for certain how many deer or birds are wounded and not recovered. Crippling losses, the term of art hunters use to describe such animals, occur and they are troubling, whether it is a deer or a pheasant or a mallard. Since bird hunters shoot at so many more birds than deer hunters shoot at deer, crippling losses occur more frequently to bird hunters. Moreover, bird hunters are routinely confronted with having to administer the coup de grace to a bird. For these reasons, even though the death of a quail or pheasant is not as consequential as the death of a deer or bear, bird hunters actually have to deal with a great deal more death than deer hunters.

This is one of the reasons why bird hunting is so closely associated with dogs. Dogs are used to find downed birds and to retrieve them. Hunting with a trained dog is the bird hunter's equivalent of the deer hunter's felt

---

*Bird hunters have to tread a fine line here. Ensuring a clean kill requires hitting the bird with many pellets since it is virtually impossible, even for a superb wing shot, to intentionally hit a bird precisely in a vital area and leave everything else untouched. The problem is that a bird that has been hit with many pellets does not exactly make the most pleasing presentation at the table. Such birds often wind up in the stock pot. The point is to hit the bird with enough shot to either break a wing or induce shock, either of which result puts the bird on the ground where it can be caught and killed.

obligation to track a deer that has been shot until it is found and, if still alive, quickly dispatched. A well-trained bird dog will mark where the bird fell and find it, often having to trail it for considerable distances. In retriever trials, dogs are scored on their ability to keep track of where two or three birds have fallen and are sent out to find each one in turn. Deer hunters talk about deer; bird hunters talk about their dogs. This reflects a tendency to think of deer in terms of specific individuals whereas birds are rarely talked about as though they were each distinct entities. It also reflects the fact that dogs are integral to the bird hunter's activity. In my interviews with dedicated deer hunters, even those who had a dog, the dog was never mentioned. By contrast, dogs were almost the first thing the bird hunters I interviewed talked about, and reference to dogs was tightly woven through their talk about hunting. It is worth recalling that Bill Crafter, who no longer has any desire to shoot animals, will nonetheless shoot a pheasant in order to honor the work of the dog. The cooperation between hunter and dog means that the chances of finding a downed bird are quite good. Suffering, at worst, is brief, and death can be delivered quickly.

Things do not always go smoothly, of course. Even the best retriever will sometimes fail to recover a bird that has been wounded, just as the skilled tracker will now and then be unable to find a wounded deer. These are demoralizing moments that hunters dread. Fishermen can boast about the size of the one that got away, but hunters generally do not talk about the deer or bird that escaped after being wounded. No one knows for sure how many game animals are wounded and not recovered. Among the hunters I interviewed, only the bird hunters talked freely about losing birds, always in the context of otherwise praising the reliability of their dog. An occasional lost bird, while an unhappy event, is accepted as "going with the territory," as Dick Board put it. By contrast, no one admitted having wounded but not recovered a deer. A few allowed that they have seen a wounded deer run by them, but I am left with the impression that this is far from a common experience. I have only seen one wounded deer in my life. The man with whom I was hunting and I waited for five minutes or so after the deer ran by us, dragging one leg, to see if the person who had shot it showed up. When no one appeared, we took up the trail; and after an hour of trailing, I finally had a clear shot at the animal, and I killed it. My companion and I were both relieved, even though neither of us bore any responsibility for the animal's plight. I recount this, not to claim special credit for myself. In fact, I think most, if not all, hunters would have done the same thing.

The desire to kill an animal cleanly or, failing that, to recover the wounded animal as quickly as possible is an ethical obligation hunters profess to take very seriously. More is involved in this than the desire to minimize an animal's suffering. Part of the ethical circle hunters forge between themselves and the animals they hunt is the obligation to honor what is killed by eating it. If the animal escapes, the ethical circle cannot be closed.

We Eat What We Kill

Russ Farina, as I have mentioned in another context, was the only hunter I interviewed who did not bring game home to eat. Neither he nor his wife likes game. Instead, he gives the animals he kills to neighbors who relish the taste of venison and wild turkey. The fact that the meat is consumed by people who appreciate wild game allows Russ to hunt with a clear conscience. For all the rest, game was celebrated as table fare. Sometimes this led to struggles, especially with children who did not like the taste of venison (or who did not like the idea of eating Bambi). Mac Braziel's daughter made plain her dislike of the game her father brought home and refused to eat it. Keith Jones's daughter was at least mildly opposed to hunting but had no qualms about eating venison. Karen DeFazio emphasizes the eat-what-you-kill imperative so strongly that her oldest two children lost interest in hunting, in part because they did not particularly like the taste of game and felt that they should really love the taste if they were to justify killing an animal.

Whatever difficulties hunters experienced at home around the eating of game, it was clear that they enjoyed eating what they had killed. Some likened the satisfaction to the satisfaction derived from eating vegetables from your own garden. It's not just that the lettuce is fresher. The fact that you grew it makes it taste better than the stuff in the supermarket. Some said they liked game because it was so lean and had no additives or hormones artificially added. Others just savored the taste. Fred Jenkins, father of a large brood, had reason to think more in terms of quantity than quality of food, but he made it plain that eating game was special. "You can't get anything like it (venison) in the market," was how he put the matter.

Almost everyone mentioned making some kind of fuss about eating game. In some households, game would be saved for special occasions. Many reported that they would have dinners once or twice a year with their hunting companions, each of whom would contribute a dish featuring a particular bird or a particular cut of venison. Men who might otherwise not set foot in the kitchen take charge of preparing game. Karl Woichek's rod and gun club puts on a game dinner each year to which members contribute a selection of the past fall's bounty, and volunteers get together to plan the menu, choose recipes, and prepare the dinner. Offerings typically include elk, moose, bear, and deer, as well as ducks, geese, and pheasants. Wives help out, but part of the specialness of the event, Karl noted, was that it was mostly men in the kitchen. Whether they are large or modest gatherings of family and friends, the fact that game is the featured item makes the event different than ordinary meals. Even in households like the Osgoods' or the Stebbinses', where game is served more often than not, certain cuts will be prepared elaborately as tribute to the animal in particular and to nature's bounty in general. Venison burger may not be rhapsodized over, but a loin will almost certainly be served with a special sauce. Several hunters noted that while they generally prefer beer, if they drink with their meal, they will usually accompany a game dinner with wine. I was offered recipes and cooking tips in the course of a number of interviews.

I noted earlier that I found little in my interviews to support the sharp distinction Kellert draws between those who hunt for food and those who hunt for sport. The reason for this is clear: hunters think of themselves as actors in an ageless drama of eat and be eaten. To kill and not eat would change the drama entirely and would necessarily place uncomfortably large emphasis on the kill itself. The eating of game—particularly those moments when the eating takes on ceremonial trappings—contextualizes the kill, after the fact, and transforms the death of an animal into a life-sustaining meal. Further, it underscores the fact that though hunters kill, they are not "killers."

There is one interesting catch to this eat-what-we-kill ethic: some hunters hunt for so-called varmits, like woodchucks and crows, or for predators, like fox or coyote, with no intention of putting them onto dinner plates. There was no pattern or consistency that I could discern in the responses to my questions about shooting critters deemed unpalatable. Rod Granger, the man whose carefully tended vegetable garden was guarded by Have-a-Heart traps, could have safely used a small-caliber rifle to "discourage" the woodchucks and rabbits. But he couldn't bring himself to shoot them. He was not quite sure why: "It just doesn't seem right," was all he could say. Perhaps he felt he had an unfair advantage. However, he no sooner said, "it just doesn't seem right," when he admitted he had shot and killed a fox the previous summer that had gotten into the pen in which he was raising several turkeys and geese. He immediately recognized the inconsistency but simply shrugged, as if to say, "you figure it out." I wish I could.

No one I interviewed deliberately hunted for foxes, coyotes, crows, or woodchucks, though several men who grew up in rural areas recalled shooting woodchucks when they were young, with the blessings of the farmers whose fields were made hazardous by the critters. Well over half said they had no interest whatever in shooting varmits. Greg Gougeon, a man in his mid-fifties who was recovering from a near-fatal motorcycle accident when I interviewed him, gave a typical response: "Hell, they have as much right to hunt as I do." Several, though, did worry about the populations of coyotes and bear growing to the point where there will be some ugly confrontations between them and humans. Al Cazminsky, the retiree who protects his bird feeders with Have-a-Heart traps, expressed exactly this fear, though he was not interested in killing bears or coyotes himself.

The dozen or so hunters who indicated that they would shoot a coyote if they chanced upon one while hunting deer or birds generally characterized coyotes as a threat to game. Though they admired the animal's cunning, they would prefer that there be fewer of them in hopes that this would mean more deer or turkey or pheasants. At the same time, however, virtually everyone I interviewed expressed delight in the prospect that wolves might make a return to New England. Several hunters mentioned hearing of reported sightings of mountain lions in Massachusetts and Vermont and indicated that they would be happy to share the woods with the big cat. In short, the hunters I interviewed did not seem to be a trigger-happy bunch. They were focused on game animals and had no interest in shooting anything that moves.

By emphasizing the clean kill, self-restraint, and eating what they kill, hunters could think of themselves as responsible participants in the natural life cycle of game animals, a life cycle that intimately connects life and death. Far from thinking of themselves as "killers," hunters see themselves as integral to the maintenance of a healthy environment. Having spent many hours talking with hunters and many more hours going over the interviews, I have no reason to doubt the sincerity of the people I interviewed, virtually all of whom claimed that they behaved responsibly, even if they might, in their youth, have done some foolish things. The question, however, remains: Is all this talk of concern for wildlife, for clean kills, and ethically driven restraint simply a way hunters mask a bad conscience and reckless behavior?

VI

BAD APPLES AND HUMAN FRAILTY

All the hunters with whom I talked gave me a version of the sporting ethic and claimed that they endeavored to meet its standards. I am sure that they do. I am also sure that, like all mere mortals, they lapse now and then. A few acknowledged a lapse or two, and several people made it clear that when they were younger they had done things that flatly violated the standards to which they now adhere. Two hunters referred to an earlier period in their lives when alcohol had them in its grip and acknowledged that they had imbibed while hunting. Many more related encounters with other hunters who were clearly violating the ethic and lacked plain common sense and decency. Dick Board stopped hunting deer after an encounter with an obviously drunk hunter. Fred Jenkins recalled going out with a couple of recent acquaintances and being horrified by their ignorance of elementary gun safety. Karen Defazio had a deer she had killed taken from her by another hunter who claimed his shot had killed the deer. She has also had bad experiences on several public hunting areas because too many of the hunters drawn to those areas are reckless or boorish or both. Critics of hunting like Joy Williams have no trouble coming up with all sorts of examples of horrible behavior—flagrant violations of game laws, casual indifference to the agony of a wounded animal, and disregard verging on contempt for others' privacy, property, and safety.

Understandably, hunters are not exactly eager to give ammunition to the critics of hunting who, like Joy Williams, are eager to paint all hunters with the broad brush of disreputability. It is hard to admit that you have behaved in ways that give credibility to your critics' charges. Critics are inclined to see the sportsman's ethic as little more than a public relations gambit, a fig leaf that can't possibly conceal the sordid stuff that hunters do. The hunters I interviewed acknowledged hunter misconduct, more often misconduct they observed, not their own, but insisted that the problem was not with hunters in general. Some hunters are jerks and slobs, they conceded, but these are a tiny minority. They insisted that the vast majority of hunters abide by the rules and thus should not be condemned for the behavior of a few yahoos. The expression "a few rotten apples" was frequently mentioned in the course of the interviews. I have no reason to doubt that the vast majority of hunters are sane, sensible, and careful. But this begs some important questions. It might help to examine the rotten-apple metaphor more closely.

A rotten apple becomes a problem only when it is in close contact with other apples. When apples are put in a barrel, one bad apple can indeed spoil the whole lot. But this is as much a "barrel problem" as it is a rotten-apple problem. The barrel in which hunters find themselves is constructed of staves that make the few rotten apples more worrisome than their small numbers would suggest. Put another way, hunting has built-in features that encourage even the most ethical and considerate hunters to occasionally depart from the straight and narrow. These features, like barrel staves, get bound together and make it hard for nonhunters to distinguish good hunters from bad. To see how this works, one must reflect on the reasons hunters hunt as well as on a feature that is so obvious it often gets overlooked or underplayed: hunting takes place in a social space that is liminal—hunters do it in the woods.

Led into Temptation

When hunters enter the woods, as we have seen, they think of themselves as entering nature, a realm where rules different from the rules of daily life apply. Not only are there different rules, the rules are self-enforced to a far larger extent than are the rules in society. Hunters, in effect, are on the honor system. They may legally take only so many birds or deer per day, but the chances of being caught with more than the legal limit of birds are really quite small. It is, of course, harder to conceal two or more deer, but it is relatively easy to conceal a dead deer and return for it the next day. Hunters are not supposed to discharge a firearm within one hundred and fifty feet of a road and five hundred feet of a building in use (homes, barns, and the like). On lightly traveled roads and in sparsely settled areas, hunters can easily ignore these laws with little risk of being caught. Game wardens are spread thinly, and even if they are alerted that a hunter has done something wrong, by the time they arrive on the scene, the miscreant will be long gone. This means, in practice, that, at best, only the most habitual and egregious wrongdoers stand a good chance of being apprehended. Hunters know that as long as no one gets hurt or the laws are not flagrantly violated, they can pretty much do as they please.

Of course, I have no way of knowing how many hunters take advantage of the fact that they can get away with something, or how often hunters cut ethical corners. Everyone has heard stories of how outhouses have been riddled with bullet holes, how hunters have mistaken a cow for a deer, and so on. For reasons I explore shortly, it is likely that these anecdotes are blown quite considerably out of proportion. Still, the fact that hunters are by and large constrained only by their consciences means that lapses are to be expected. The probability of lapses is no doubt raised a notch because hunters, in addition to being largely self-policed, are anonymous. Only in the most insular and tightly knit communities, of which there are fewer and fewer, do hunters and nonhunters know one another. I spoke of this earlier in terms of the depletion of rituals that linked hunters to nonhunters. As hunters and nonhunters become socially separated from one another, hunters have become something of an invisible minority. Peter Hanks, the retired elementary school principal, concealed the fact that he hunted and enjoyed target shooting, lest the parents of his students become alarmed. Game used to be shared among neighbors, but most hunters report sharing game only among family members and within circles of hunting buddies.* Even Peter Hanks's neighbor of many years did not know that he was a hunter.

Anonymity increases with the distance hunters travel to their hunting

---

*In rural areas, hunters will wear a blaze orange hat or jacket even when they are not hunting, and gun racks in the back of pickup trucks are common. But hunters in cities and suburbs are far less likely to wear hunting garb or otherwise call attention to the fact that they hunt.

coverts. Except for a comparative handful of hunters who can hunt on their own land and the slightly higher number of hunters who can hunt on public or private land in their immediate vicinity, most hunters have to travel an hour or two to get to a place to hunt. Those who go to the same place year in and year out no doubt cultivate relationships with the landowners on whose land they hunt. But many hunters cruise the back roads looking for likely spots, and when they spot an area that looks promising, they park and commence hunting.* This means that neither hunter nor landowner need know one another. It seems reasonable to assume that we tend to be more thoughtful with and considerate of the property of those whom we know. If this is so, then anonymity certainly must add something to the forces that weaken a hunter's sense of constraint.

Hunters, with few exceptions, are "on vacation" when they go hunting: they have checked out temporarily from the workaday world. Being on vacation, as anyone who has ever worked at a resort or lived in an area known as a vacation destination can attest, does not exactly bring out the best behavior among vacationers. Vacation is defined as a time when it is okay to relax inhibitions, which is no doubt one of the reasons people so often quip that they have to return to work in order to recover from vacationing. Many of the traditions associated with hunting camps involve drinking and large, heavy meals. Even though these traditions appear to be on the wane (Miller 1992), there are no doubt still plenty of examples of hunters eating and drinking to excess. This does not mean that hunters are commonly drunk while they hunt, though a few of the hunters I interviewed had encountered a drunk hunter on occasion; but those who experience the reduction of inhibitions associated with being on vacation certainly must feel the added temptation to bend the rules.

There is one more feature of contemporary hunting that must be factored in. Most hunters have to reconcile themselves to a few outings a year. Seasons are short, and bag limits are such that most hunters have to pack their hunting into fewer than twenty days a year. Cal Jones, the carpenter who tries

*There is an important regional aspect to this. In New England, dating to the early colonial period, land not explicitly posted is presumed to be open to hunting. Hunters are encouraged, out of courtesy, to seek permission of the landowner, but this is not required. In the South and the West, access to private land is by permission only. Increasingly, landowners are charging fees for access. Antihunting groups in New England have sporadically tried to pass "by permission only" laws in an effort to make hunting more difficult, but so far, they have had only limited success. I return to this effort briefly in the next chapter.

to arrange getting laid off in the fall so that he can scout and then hunt nearly every day, is the exception. Fred Jenkins, the stonemason, is far more typical. Fred takes a week off to hunt, and in a good year, he might add a day trip or two. Greg Gougeon hunted frequently the year before I interviewed him because he had been laid off. Now that he's back working, he was resigned to hunting for a day or two at most—new to the job, he had little vacation time coming and could not afford to take unpaid time off. Karl Woichek, like the other teachers I interviewed, could hunt only on weekends. And so it went. Family and job, coupled with the need to travel to and from hunting areas, produce a situation in which hunting opportunities are compressed into a few days or a few weeks. Karl Woichek, as he told me about how difficult it was for him to get away, noted ruefully that he spent far more time thinking about hunting than he spent actually hunting.

Compressing hunting into a week or two, or into a string of a half-dozen weekends or so, also must add to the intensification of what begins as an intense activity. Under these circumstances, the desire to get a deer or bird is surely heightened, not because hunters are trying to prove their manhood (or to prove that they are just as good as "the boys"), but because they have spent so much time and energy (and money, let us not forget) preparing for this outing. If a law stands in the way, or if one's own deeply held ethical convictions collide with an opportunity to get a shot, it is going to be tempting, to say the least, to do something one knows is wrong but which, under the circumstances, is likely to be harmless and to go unnoticed.

Since there is no good evidence on which I can draw to make anything more than a wild guess about how hunters actually behave in the field, let me share some anecdotes from my own experience before revisiting the "bad-apple" characterization.

I was in a duck blind and had been waiting for months for this morning. My companion's springer spaniel was sitting patiently, waiting for the command "Fetch." I could hear ducks flying back and forth over our decoys, but it was too dark to make them out. At first light, shots rang out just upstream from our blind, and every duck in the county was instantly in the air and headed for safety. That was it for the dawn's hunt. The sun rose and quickly burned off the fog that had hung along the river, revealing the sort of cloudless sky that duck hunters hate. We saw a few ducks, but they were well out of range and showed no interest in our decoys. After an hour or so, two mallards turned to consider the decoy set, and as they passed at sixty yards, I shot, even though I knew that at that range I'd cripple, not kill. Fortunately, I missed cleanly but, truth is, I should not have shot at all and I knew it. And I shot anyway.

I was talking about shotgun range and accuracy with a fellow with whom I'd been shooting skeet. He said that he never takes a shot at a deer much beyond sixty yards or so—his own personal rule. No sooner had he said this than he went on to say that just last year he saw a really fine buck ninety or maybe even a hundred yards away. He shot—and missed. "I knew the odds were bad, but it had such a huge rack I just couldn't resist."

Years ago, when I first resumed bird hunting after many years of not hunting at all, I was having a devil of a time. I was new to the area and knew few places to hunt, and I was without a bird dog. As a result, I saw very few pheasants, try as I might, and the few I did manage to flush I missed. One morning, as I finally began to figure out the habitat and began to find birds, I missed two pheasants in rapid succession and was fit to be tied. The echo from my second shot had barely ended when a crow flew overhead. Without thinking, I raised my gun to my shoulder and fired. The crow fell dead to the ground, and I felt completely ashamed. I should have eaten crow, literally. Instead, I sheepishly buried it, as much to conceal my rotten impulse as to pay my respects to the hapless crow.

One late afternoon, many years after my early frustrations, I was heading for the car after a couple of hours of good hunting—which meant that my dog had worked exceptionally well and I had shot decently. As I walked, my mind left hunting and turned to the evening ahead and classes for which I had to prepare. I was going over some obscure matter when a cock pheasant flushed well ahead of me and sailed into the brush that marked the beginning of someone's backyard. My dog saw the bird, too, and regarded it carefully as it settled into the thicket. She held fast, and at that moment I should have given the command she least likes to hear: "We're going in, Stella." Hunt's over. Instead, I heard myself say, sotto voce, as if keeping the secret from myself, "Find the bird." Stella, a strong and almost obsessive German shorthaired pointer, was off like a rocket, heading straight for the spot she last saw the bird. I stayed where I was, well beyond the five-hundred-foot distance hunters must maintain between themselves and an occupied building. I was mostly interested in watching the dog and bird play, if you will, cat and mouse, but I did reload my shotgun just in case the bird flushed back toward me. The spell was broken by the irate hollering of the woman whose backyard Stella had taken over. I was plainly visible in my blaze orange cap and vest. I knew I was not going to shoot toward her house, but the home owner had no way to know that. As far as she was concerned, I was hunting in her backyard. I whistled for Stella to return, but she was oblivious to the shouting and whistling. I would have to go get her and no doubt further terrify and enrage the woman, or wait helplessly until Stella put the bird to flight. Remember—Stella was a pointing dog, and a good one at that. So the woman and I were treated to a classic scene of a dog on point that lasted three or four agonizing minutes, until the pheasant's nerves gave out and it took off. Needless to say, the woman did not share my delight in pointing dogs. A happy Stella came to my side for praise for a job well done. The woman glared as I gave Stella a desultory pat, and we resumed our trek to the car at a brisk pace, this time with Stella on leash.

In thirty years of hunting, I can recall only one or two other incidents, in addition to those just described, in which I did something I knew was ill advised. None of these lapses resulted in personal injury or property damage. The point of sharing these anecdotes is as simple as it is important: while it is no doubt true that there are very few thoroughly rotten apples among the ranks of hunters, there is a little rotten apple in most if not all hunters. The context makes it all but certain that even the most stalwart and conscientious hunter will fall from grace once in a while. Motive and opportunity, to borrow a page from criminal investigations, are almost continuously present. The wonder is that hunters are as well behaved as they are.

I have no way of knowing whether my record is better or worse than those of other hunters, and there is no way of knowing how many homeowners are frightened or irritated by hunters. I am tempted to conclude that the large majority of hunters are sensible and considerate most of the time, but in the absence of trustworthy information, I will have to leave this to each reader's judgment. One thing, though, is known with considerable precision, and it bears directly on the question of how responsible hunters are: hunting accidents have been steadily declining for several decades, and hunting now ranks among the safest of all outdoor recreations. In 1999 (the most recent statistics available as this is being written), there were 67 hunting-related fatalities, 21 of which were self-inflicted, and 661 nonfatal accidents, 202 of which were self-inflicted. Twenty of the fatalities were caused by hunters under the age of nineteen. Alcohol was implicated in only 2 fatalities (and 3 nonfatal accidents). Careless handling of firearms caused 10 of the fatalities. It is also important, for reasons that will become clear shortly, to note that almost all of the fatalities are inflicted on hunters. In recent years, at least, there has rarely been more than 1 or, at most, 2 fatalities in which a hunter has shot a nonhunter. In most years, no nonhunters are killed.*

Though each of these accidents represents a serious tragedy, the fact is that millions of hunters, each hunting for at least several days over the course of a hunting season, make a negligible contribution to the number of national accidental deaths. In 1998 (the most recent data available), there were 51,400 non–motor vehicle accidental deaths, 93 of which were attributable to hunting. In other words, hunting was responsible for .00002 percent of the accidental deaths in 1998. Moreover, for all the talk of drunk hunters, alcohol appears as a factor in hardly any of the accidents, fatal or nonfatal, and rank carelessness is also not a major issue. Why, then, is the reputation of hunters so poor? Why do so many people fear for their lives when hunting season opens?

When Hunters and Nonhunters Meet

Even under the best of circumstances, hunters appear menacing. After all, they are armed, and their weapons are carried in plain sight. Their blaze orange vests and jackets make them seem larger than life, and when they don camouflage and face paint (archers and turkey hunters), they look as if they just stepped out of a war movie. Hunters also materialize without warning. The homeowner whom Stella and I upset could not have seen me coming and had no way of knowing if I was about to sneak into her backyard to shoot the pheasant. After all, hunters are furtive, they appear and disappear such that residents in areas where hunting is common have no way of knowing who is out there, how far away they are, or how safety-conscious they might be. The stage is unavoidably set for breeding apprehension. To make matters worse, chance contacts with hunters almost always find hunters looking their worst. Likely to be unshaven, sweaty after a day of being on the move, even the most refined and urbane hunter will

---

*These data are compiled annually by the International Hunter Education Association and are available on their website: <www.ihea.com>. The information regarding nonhunter fatalities was offered in personal conversation with Dr. David Knotts, president of the IHEA.

look scruffy and unkempt. This is hardly the sort of appearance that inspires confidence.

Relationships with landowners are, nevertheless, established and ease landowners' misgivings, at least for those hunters whom they know. Many of my respondents indicated that they had, over the years, forged friendships with some landowners, typically farmers, and felt warmly welcomed when they showed up on the eve of opening day. Karen DeFazio returns each fall to a farm in west central Massachusetts that she first hunted as a child accompanying her father. That farm, in a sense, had been in her family for nearly fifty years. Zane Zacharias hunts on properties his grandfather and father had hunted, and he knew the landowners as friends and neighbors. This said, it is also the case that few hunters can rely exclusively on such close relationships with landowners, unless, that is, they are content with hunting a few coverts over and over again. Moreover, the days when land tenure was measured in generations is passing very quickly. Even if farms remain undeveloped, new owners may be unwilling to honor the friendly arrangements with a group of hunters that the previous owner had established.

Greg Gougeon has seen almost all of the farmland he began hunting twenty-five years ago change hands. Much of it has become subdivisions that have formed an ever-expanding circle around Springfield, his place of residence. What remains open has largely become off-limits as the farm generation retires or dies off. Greg reported that one landowner declined renewing permission for Greg to hunt because the landowner had been pressured by people in the new subdivisions nearby to close his land to hunting. The landowner was apologetic but felt he had no choice but to be a good neighbor and honor the wishes of his new neighbors.*

Greg has adjusted to this change by moving further west to do his hunting, but he barely stays ahead of the sprawl. This means that he is acutely aware of the difficulties of hunting in rather close proximity to suburbanites and exurbanites. He has had several encounters with people who became terrified when they looked out their bay window and saw Greg

---

*Interestingly, farmers report a similar sort of thing. People are eager to move to the country to seek the presumed comforts of a more rural way of life. Once in the country, with working farms as their neighbors, they discover that farms come with lots of noise—and noise not conveniently confined to nine-to-five—and lots of unpleasant odors. Complaints abound, and exasperation rises on all sides.

emerge from the woods that grew to the edge of their backyard. Sometimes, he said, he will try to engage them in conversation, explain what he is up to, and reassure them that they have nothing to fear. However, he has been doing less and less of this. "They just don't understand. They hear a gun go off a quarter mile away, and think they're in mortal danger," he observed. He can understand their concern but not their eagerness to jump to dire conclusions. A number of other hunters reported similar sentiments verging on consensus: the nonhunting public has an irrational fear of hunting.

The fear that hunters inspire may indeed be irrational; certainly, from the point of view of the statistics on hunting accidents, fear seems unwarranted or at least vastly exaggerated. Nevertheless, the contexts in which non-hunters encounter hunters cannot help but fuel such fears. Greg is not an imposing fellow. But if I were sitting in my breakfast nook lingering over my last cup of coffee for the morning and I spotted him (remember that the mandatory blaze orange outfit he wears while deer hunting makes him stand out like a roman candle)* moving through the woods just beyond my property line, I would be alarmed, too. How could I possibly know if he was the calm sensible guy I interviewed or some bozo who never saw a safety precaution he took seriously?

I can say from personal experience that it is far from an easy or pleasant task to engage a homeowner or landowner under such circumstances. To be sure, some hunters are not all that articulate to begin with, and they may be ill at ease, even in uncharged situations, talking with strangers, especially strangers who seem better educated and more sophisticated. But this is not my problem, and it is not a problem for the majority of the men and women I interviewed. The context overwhelms personal qualities. When I have a gun in my hands and am in blaze orange, I cease being a professor at one of the nation's premier liberal arts colleges. I am a *hunter*. There have been some exceptions, but they've been few. Once, I was driving to a bird covert I had hunted for years and came across a fellow putting No Trespassing signs up all along the covert I was intending to hunt right then. At first my heart sank, and I was about to drive on. But then it dawned on me that I would never have an easier time introducing myself to the

---

*It should be noted that mandatory blaze orange has contributed significantly to the reduction of hunting accidents. In the handful of states that do not require blaze orange, fatalities and serious injuries are almost entirely confined to hunters who were not wearing blaze orange. Massachusetts was one of the first states to require blaze orange.

landowner, and who knows, he might be posting his land to keep deer hunters out but would have no problem with a bird hunter. So I pulled over and approached the guy. I introduced myself and extended my hand. As I was explaining myself, his eye caught the college parking sticker on the rear window of my vehicle. His nephew had attended the college (no, I hadn't had him in class), and he held the place in high regard. I have hunted that property for many years now, stopping by when I see the man's car in the drive, to say hello and exchange pleasantries. I wish I could report that this was a common occurrence. In my experience, it is not. The bird hunters I interviewed indicated that they thought most landowner antagonism was directed at deer hunters and that they stood a better chance of getting permission to hunt land that was posted if they could make clear that they were not deer hunters. They also noted that it wasn't all that easy to do.

There are some important differences between bird hunting and deer hunting that may produce some differences in the two sets of hunters that bear on public acceptance. Deer hunting produces many more accidents, fatal and nonfatal, than other types of hunting. Indeed, in 1999 upland bird hunting (pheasant, quail, grouse, and woodcock) caused only one fatality, compared to thirty-six fatalities attributed to deer hunting. The bird-shot used in upland hunting is small and loses energy rapidly. To receive a fatal injury, a hunter would have to be at close range (and thus quite visible) and be hit by nearly all of the pellets discharged. Moreover, birds, are shot at while they are in flight, which means that the shot is aimed upward, generally well above the heads of fellow hunters (assuming they are on more or less level ground). The chances of being mistaken for a bird are obviously not large. By contrast, deer are shot with bullets, slugs, or buckshot which are much larger than birdshot and can cause mortal injury over much longer distances than birdshot. Longer distances mean more chances that a person could be mistaken for a deer. Moreover, deer are on the ground, and their stature is very nearly that of an average human—a shot to the chest of a deer would also hit the chest or stomach of a human. The result is more injuries and more injuries that prove fatal.

There is also a difference that arises from the difference between large and small game. Earlier, in another context, I called attention to the fact that killing a deer is a more overpowering experience than killing a pheasant (or a squirrel or rabbit). The difference in size matters, but it is not sheer size that matters most: it is what size means. There is no record grouse or pheasant. Small game is, well, small, and there is hardly perceptible variation within species. By the fall, the grouse or rabbit born in the spring is about as large if not as large as its parents. The same, of course, is not true for deer (and other big game). There are recordbook deer, both by body weight and by antler measurements. There isn't, I would assert, a deer hunter alive who hasn't dreamed of bringing home a large buck with an impressive set of antlers. No one, certainly no one I interviewed and, for that matter, no one I've known, goes bird hunting having dreamed of bringing home the largest woodcock. Thus, there is, to the best of my knowledge, no equivalent in upland bird hunting to so-called buck fever.* Deer hunting awakens passions that many other types of hunting do not. Passion and safety can make for unsteady companions.

It is hard to say how many landowners know or care about such distinctions. It is probably fair to say that people who have lived their lives among hunters have at least an intuitive sense of the different risks deer and small-game hunting pose, but the point is quickly becoming moot, at least in states like Massachusetts, where the character of the rural population is undergoing a sea change. As likely as not, the residents of the homes in rural Massachusetts are recent arrivals from the city, and they have had little or no experience with hunting or hunters. People move to the country, whether for their primary or second home, for the presumed peace and tranquillity the country offers. Hunters disrupt this idyll. Even if no trespass occurs, the urban transplants lose the sense of solitude they came to the country to find. Again, there are differences between small-game hunting and deer hunting, but both types of hunting have their own sorts of irritants. Massachusetts deer season, as I have noted, is relatively short—twelve days for the most popular method of deer hunting (shotgun). By contrast, upland bird season lasts roughly six weeks, and the season for cottontail rabbits is over four months longs. The short deer season means, among other things, that deer hunters are out in force, most heavily, of course, on opening day. Since upland game

---

*Being overwhelmed by excitement to the point of loss of control may occur in turkey hunting and in waterfowl hunting, both of which entail lying in wait, sometimes for an hour or more. When a bird finally appears, the temptation to shoot can be overpowering and does cause some hunters (or most hunters once in a while) to take ill-advised shots, generally shots at excessive range. Duck and goose hunters disparagingly refer to this as "sky busting."

hunters have a much longer season, they do not descend upon the countryside in hordes the way deer hunters do. Though their presence at any moment in time may weigh less heavily than the presence of deer hunters, the fact is that upland game hunters will be a presence among residents of rural areas, who will have to share "their" woodlands and fields, for several months.

This presence would be less an irritant were it not for guns. Most urban transplants have as little acquaintance with guns as they have with hunting. The sound of a shotgun carries much farther than the projectiles it sends forth. To someone unfamiliar with guns and unaccustomed to them being fired, opening day can seem like a small war. Shots, several to a volley, are not uncommon in the first few hours of the opening of pheasant, duck, and deer seasons. To the uninitiated, the peril must seem extreme, especially since the different risks posed by birdshot and buckshot are likely less impressive than the fact that both make the same noise. Though there are still some areas in southern New England where hunters largely have coverts to themselves, places where there are few residents to irritate, for the most part, hunting goes on within earshot of nonhunters. This means that even if hunters are perfectly behaved, they are going to upset some people. And when they are not perfectly behaved, as I was not on that day I angered a woman by letting my dog roust a pheasant from her backyard, the chances are good that someone will witness the lapse. There is no doubt in my mind that the woman I offended, whatever she may have thought about hunters before that encounter, came away with a very uncharitable view of hunters.

The nature of hunting creates conditions that produce occasional departures from the standards of sportsmanship and good sense that virtually every hunter believes in. So, while only a very small number of hunters are unreconstructed jerks, a goodly number of hunters do something once in a while that they know they should not do and are not likely to repeat. If all hunting took place in vast empty stretches, no one would be the wiser. Population growth and the attractiveness of rural areas being what they are, however, hunters increasingly hunt in areas where their behavior can be observed by nonhunters. While more true of southern New England than of many other parts of the country, the trend nationwide is clearly pointing in the direction southern New England has taken. Hunters may not yet be in a fish bowl, but they no longer have the luxury of having the woods to themselves.

Of course, most nonhunters have no firsthand encounters with hunting. If they hear gunshots, they believe themselves endangered. Their insulation from hunting and isolation from hunters does not, however, mean that they are necessarily indifferent to hunting or have open minds about the character of hunters. If the interactions between nonhunters and hunters in rural areas puts nonhunters' teeth on edge, the isolation of urbanites from hunting and hunters produces precisely the social context in which negative stereotypes are most likely to flourish.

Did You Hear about That Hunter Who . . .

In the absence of direct contact with hunters, the bulk of Americans have no basis but anecdotes on which to judge hunters. For many decades this did not matter much. Regard for wildlife was low, hunting was securely embedded in the national narrative as one of the key ingredients of what made Americans free and independent, and hunting accidents, which were quite frequent until they began to decline sharply in the 1960s, inflicted a toll only on hunters themselves. But things change. For reasons I need not delve into here, attitudes toward wildlife began in the 1970s to turn from fear or indifference to appreciation and solicitude (Duda et al. 1998; Gray 1993). Growing concern for the environment in general, and for endangered species in particular, helped speed this transformation, so that by the late twentieth century even wildlife species such as the wolf, which had long been demonized, became centerpieces in the popular environmental movement. At the same time, for completely unrelated reasons, a rapid rise in urban violence, especially gun violence, as well as a string of attempted and successful assassinations of prominent political and cultural figures, broke the nation's complacency with respect to guns. The close association of hunting and guns implicitly aligned hunters with "gun nuts." Finally, urban sprawl and rural gentrification brought more Americans into uneasy contact with hunters.

Into this combustible mix a lit match was dropped. In November 1989, on the outskirts of Bangor, Maine, a young woman, mother of several children, was in her backyard. She and her husband and children had recently moved into a newly built house at the end of a cul-de-sac that placed their home into what had been an unbroken stretch of woods. It was a chilly morning, so she was wearing white mittens.* A deer hunter mistook the woman for a deer, and he fired at her, killing her instantly.

News of the tragedy spread quickly. Hunting magazines and outdoor columnists aired the mixed views this horrible incident gave rise to. Many faulted the hunter for having violated the cardinal rule of shooting: see your intended target clearly and make sure there is nothing behind your target that could be harmed before you squeeze the trigger. Obviously, the hunter had not done this. Others rose to his defense, noting that though the hunter was familiar with the area in which he was hunting, the house was a new home that had been tucked far into the woods. The mistake was an "honest one": he wasn't drunk, he wasn't blazing away indiscriminately, and he had not tried to escape the consequences of his act. He was, in fact, a responsible member of the community. A hearing subsequently found the hunter innocent of any wrongdoing—it was judged an accident to which no culpability could be attached.

This tragedy is still talked about by people who feel besieged when deer season rolls around. With every recounting, the grisly story is gradually transformed from an exceedingly rare, even freakish, event into an example of general risk. I do not mean in the least to make light of Karen Wood's death, but it is important to remember that in the years since that fatal mishap, there have been very few nonhunters shot by hunters in the country as a whole. Each life lost is tragic, but the fatalities do not even begin to rise to a level where one could reasonably argue that hunting constitutes a general menace. And yet, despite the rarity of such tragedies, the view that hunters pose a threat to the public persists. In effect, the incident in Maine formed the basis of what might best be thought of as an "urban legend."

A distinct social process is at work when a false or distorted view achieves

---

*There is conflicting testimony in the voluminous records of this case. The hunting companion of the hunter who shot Karen Wood rushed to his partner's and Wood's aid did not see any white mittens. There is also conflicting testimony about whether she was hanging clothes. Shots had been fired moments before Mrs. Wood was fatally shot, and the companion reported hearing a woman shout, "Don't shoot," or words to that effect. She was found lying near the property line and the beginning of the forest, yielding speculation that she may have gone to the edge of her yard and shouted to alert whoever was shooting that they were close to houses. Whether it was her movement or the contested white mittens that drew the hunter's fire, we will never know. I am grateful to the Maine Department of Inland Fisheries and Wildlife for making the file on this tragedy available to me.

the status of a self-evident truth. In his book, *The Culture of Fear: Why Americans Are Afraid of the Wrong Things* (1999), sociologist Barry Glasser examines how it is that we come to exaggerate the risks involved in our daily lives. Although he does not mention fear of hunters, when he describes how our anxieties over crime, road rage, and a wide range of other real or imagined perils can be raised to a fever pitch, he could just as well be talking about the nonhunter's perceptions of hunters. Key to the exaggeration of risk is the way events get depicted. Because few of us have any direct link to events, we are necessarily dependent upon news organizations of one sort or another to tell us what is going on in the world. These organizations vie for our attention, and one of their understandable gambits is to sensationalize or otherwise blow out of proportion this or that event. This strategy is compounded by advocacy groups who, however well-intentioned, are committed to portraying their particular cause in terms as urgent, even alarmist, as possible. Under these circumstances, people can easily be led to think the world is coming apart. So animal activists proclaim we are in the midst of an unprecedented "war on wildlife."* The "body count" of doves, deer, and other game animals is in the many millions annually. A naïve person might well conclude that wildlife is being pushed to the brink. But, of course, the very opposite, as I have shown, is the case. If hunters are waging a war on wildlife, they are clearly losing—hunted species of wildlife, with very few exceptions, are thriving.

In the absence of hard evidence or the information needed to put an event into context, it is easy to conclude that hunters create havoc. Thus, the farther removed from hunters and hunting the general public has become, the harder it is for them to judge the risks to which hunting subjects nonhunters. The story of the accident in Maine is recalled whenever there is a report of a hunting-related mishap, as if the tragedy in Maine was commonplace. After all, it seems that this should be so. The news is filled with reports of gun violence; hunters have guns and go out to kill "for sport," that is, they kill animals for the hell of it; and hunters are suspiciously disreputable.** These widespread impressions make it easy to see how nonhunters can conclude

---

*The phrase is Wayne Pacelle's. Pacelle is one of the prominent spokespersons for the Humane Society of the United States' campaign to end hunting.

**Donna Minnis (1997) has compiled a very useful "typology of beliefs" that constitute the basis for opposition to hunting. I have emphasized here only a few of the more consistently expressed themes.

that hunters are dangerous. On the basis of widely shared assumptions about hunting and hunters, the conclusion seems plausible.

In a similar vein, social scientists have begun to examine so-called urban legends—unfounded rumors that attain the standing of self-evident truth. Patricia Turner, in her book *I Heard It through the Grapevine* (1993), examines the process by which unfounded rumors are kept alive within the black community. One such rumor has it that black people are being used as guinea pigs by the medical establishment. The key is that the rumor has a shred of plausibility that then gets magnified and generalized all out of proportion. Blacks have been used as guinea pigs without their knowledge or consent: the infamous Tuskegee experiment in which blacks who had contracted syphilis had the diagnosis withheld and were not given antibiotics so that the doctors could study the course of the disease. There may well be other instances of such abuse, but there is no basis for the rumor that such practices are widespread or that they constitute a concerted effort to destroy black people. Still the rumor persists, as do rumors about fast food chains putting harmful ingredients into the foods they sell to black customers. If a person expresses doubts about the truth of one or more of these rumors, as Turner shows, he or she is quickly reminded of all the bad things whites have done to blacks over centuries; case closed.

Jerry Lembcke (1998), a sociologist and veteran of the Vietnam War, exhaustively researched the allegation that returning soldiers were spat upon by antiwar protestors. He could find no corroborated evidence of such an incident, never mind evidence for the rash of such incidents that rumor alleges. A protestor may have spat upon a returning vet—there can be no doubt that feelings were at a fever pitch, and there was, for all the love-ins of the time, little love lost between the protestors and many soldiers. In this context, it is hard to believe that no insults, obscene gestures, or saliva got exchanged. It is clear, though, that spitting in particular was, at the most, exceedingly rare and likely even a nonevent. As do blacks, Vietnam vets feel beleaguered or misunderstood or out of favor. Rumors such as being spat upon help to create a shared belief about how insidious forces are at work to make their lives miserable. Anti-Semites, racists, neo-Nazis, and extremists of all stripes also function in this way, marshaling shreds of "evidence" and weaving them together in an internally consistent analysis that is a caricature of reality (Ezekiel 1995).

I am not suggesting that people residing in areas frequented by hunters are kissing cousins to paranoids because they grow anxious when hunting season draws near. Nor am I claiming that nonhunters in general, are gullibly willing to believe any derogatory claim made about hunters. There are risks whenever a firearm is discharged, and even though the risk of being injured or killed is minuscule for hunters and even tinier for nonhunters, many people are understandably unhappy having even a tiny risk thrust upon them. Even if one accepts the fact that the vast majority of hunters are sensible and cautious and pose no danger to anyone, how can one be sure that the person hunting in the woods behind one's property isn't one of the rare bozos who shouldn't be trusted with a peashooter, much less a firearm? In the context of just this sort of uncertainty, the rare accident looms larger than life and gives people a sense of vulnerability.

As far as the woman I angered when I sent my dog into her backyard after a pheasant was concerned, I was representative of all hunters. The fact that I posed no risk whatever is beside the point. She felt intruded upon and in danger, and I knew I should have had my dog under control. I wrote a note of apology to the lady that evening, but I suspect that did not change her low opinion of me or of hunters in general. Moreover, for reasons we have already explored, I will bet that she has related that example of my stupidity to friends near and far who all take that episode to be confirmation of how little hunters are to be trusted, much less tolerated. I was a "slob hunter" for ten minutes, though I broke no laws, took no shot, and was contrite. The way even isolated incidents get magnified by the retelling, even if most hunters slip from the straight and narrow only twice in the course of their hunting careers, this is sufficient to cast the whole lot of hunters in an unfavorable light. Add to the mix the handful of inveterate "slobs," and it is easy to see why the general public has such low regard for hunters.

Interestingly, even hunters themselves are not all that trusting of other hunters. I have already noted that a number of the hunters I interviewed had unflattering things to say about their fellow hunters. Hardly anyone with whom I talked would be willing to accompany a complete stranger on a hunt. They would first want to know something about the person's temperament, judgment, and competency with a firearm. There is no reason to presume that the stranger will behave sensibly and safely. If hunters are cautious about the company they keep while hunting, it is small wonder that nonhunters are concerned. Every hunter I interviewed either hunts alone or with a small group of friends or kin that is quite stable over long periods of time. There is also a marked preference for returning to the same areas year after year. In part this is a function of nostalgia, but it is also a matter of safety. Familiarity with an area translates into knowing where others in your party are so that you come to acquire almost a sixth sense of what shots are safe. Similarly, people who have hunted together for years come to know one another's habits—how fast they move through the woods, what sorts of shots they will take or pass up, how long they will remain in one place before moving on—and this knowledge makes it much easier to anticipate where one's companions will be. No one likes to encounter a stranger in such a situation, not because hunters are unfriendly or possessive of "their" hunting area, but because sharing the woods with strangers introduces an element of unpredictability. Hunters are thus caught in a dilemma. They are defensive when nonhunters express fear of hunters, and yet they understand full well that there are reasons to be concerned. It is frustrating, to say the least, to behave responsibly and have that trumped by an accident that happened in another state fifteen years ago.

Even though the sportsman's ethic makes it abundantly clear that hunting is a privilege that turns on private landowners' good will (and virtually every state's booklet of the hunting laws that accompanies each license emphasizes the importance of hunters cultivating good relationships with landowners), many hunters talk as if hunting is a God-given right. If directly asked, of course they acknowledge that their "right" to the land is not really a right, but it is clear, more from the *way* hunters talk about the land than what they *actually say,* that they view restrictions on their access to the land as unfair and unreasonable. Anyone who has spent any time on the back roads of Massachusetts (or Vermont or New Hampshire) has seen Posted or No Hunting signs peppered with birdshot or ventilated by a shotgun slug, a clear expression of the resentment hunters feel about land that is off-limits. By contrast, one can search in vain to find a Safety Zone sign defaced. Safety Zone signs warn hunters of occupied buildings nearby and establish a perimeter beyond which shooting is safe. Hunters appreciate such warnings. Indeed, some rod and gun clubs offer to put up Safety Zone signs for landowners in hopes of averting some incident that could result in injury or irritation that causes the landowner to post the land.

Only a handful of the hunters I interviewed admitted to having engaged in behavior of which they are not proud, and those who did all spoke of their ethical failings and recklessness as youthful indiscretions they would no longer dream of committing. Andy Felter shot his mother's cat and had to relinquish his gun for a time. Mac Brazeile was caught by his father shooting sparrows with his BB gun and was made to clean and eat the birds as a way of driving home the lesson that wanton killing is bad. Russ Farina did not go into specifics but made it clear that he and his teenaged companions "did some pretty stupid things." I will bet he and his buddies chose some pretty inappropriate targets on which they demonstrated their marksmanship, both animate and inanimate. Robert Swipe, the man who grew up in a state home for the indigent, began shooting with no instruction whatever and shuddered when he reflected on how dangerous he was, though he didn't know so at the time. He would shoot at makeshift targets without any sense of where his bullets would be going and was himself shot in the leg by an equally inept young companion. As he put it, "It was a miracle we didn't kill someone."

Such admissions as these were infrequent, but I will bet that there are few hunters alive who have never given in to the impulse to shoot at something inappropriate or have never taken a bad shot, one more likely to wound than kill outright, or one that was unsafe. And while it is no doubt true that most of this occurs when hunters are young and "feeling their oats," just as it is true that killing game is more important to the young hunter than to the seasoned hunter, it is also likely that even experienced hunters occasionally let down their guard and do something stupid or unethical.

I can think of no reason to single hunters out in this regard. Normally conscientious hikers will occasionally toss a Power Bar wrapper onto the trail or pick a lovely wildflower that catches his or her fancy; bird-watchers are no doubt tempted once in a while and get too close to the object of their fascination or venture into a sensitive nesting area. Despite the signs and warnings, we feed wild animals; we run stop signs, routinely exceed the legal speed limit, and at least occasionally drive with two or more drinks under our belts. We are, in short, a refractory species. Hunters are no different—except that hunters carry weapons that can maim and kill. This single fact throws a spotlight on hunters that is rarely if ever thrown on hikers, rock climbers, bird-watchers, and others whose use of nature is regarded as "passive."

For their part, as I've suggested, hunters have in fact done a lot to clean up their act in recent years, the horror stories notwithstanding. There will always be horror stories; the point is that they are not even remotely close to reflecting common experiences. Hunter education required of all new hunters has, as I've noted, dramatically improved hunter safety and has almost certainly contributed to a wider acceptance of the sporting ethic. Changing patterns of recruitment to hunting have also contributed to improving the behavior of hunters. Traditionally, most hunters were recruited the way Zane and Jack, the two young men from western Massachusetts, and Tommy Worthington were: they began tagging along with older male relatives, typically fathers and uncles, and learned how to hunt by observing their respected elders. However heartwarming this might be in an era when so-called family values are alleged to be in retreat, it must be said that lots of bad habits got handed down from father to son. Fathers do not always know best—and even if they do know what is best, their behavior does not always and faithfully spring from this knowledge. Kids can see as well as hear.* Hunter education gives youngsters another model of behavior to follow, a model that is almost certainly better than that which most get at the side of an older male.

In recent decades, as I noted in chapter 1, the traditional father-son recruitment to hunting has been declining. This means that more and more of the nation's hunters are coming to the sport *as adults*. Adults do not bring with them the same "trigger-itch," to use Aldo Leopold's term, that seems to grip so many adolescent boys, perhaps because they have less to prove to themselves. Whether we are talking about guns or cars, something quite dramatic occurs around twenty-one years of age. Twenty-one does seem to mark the point at which, in our society at least, risky behavior tails off and prudence gains a decisive foothold. Moreover, as Mary Zeiss Stange (Stange and Oyster 2000) has so insightfully argued, more and more of the adult recruits to hunting are women, who bring to hunting sensibilities that are different from those our society inculcates in boys. It is not that women are genetically programmed to be more ethical or more self-controlled— the kinder, gentler sex of Victorian imagination. Women are far less likely than men to assume that they know all they need to know about guns (or bows) and game animals. Firearms instructors and outdoor educators are virtually unanimous in noting how much less guarded and defensive women are than men about their naïveté when it comes to guns and hunting. Somehow, men feel as if they should know this stuff intuitively. Women do not.**

---

*It would be well to remember in this connection just how fine a job parents traditionally did in educating their offspring about sex.

**I look at this distinction from a slightly different angle in the next chapter when I explore how hunters feel about encouraging their daughters to hunt. It is too bad that there are no studies that have been done comparing traditionally recruited and nontraditionally recruited hunters, both male and female. Now that more hunters are nontraditionally recruited, such a study would be relatively easy to carry out.

This important difference aside, both adult men and women bring to hunting already formed and to some degree tested ethical sensibilities. Starting fresh, they are also less likely to bring bad habits and misinformation with them when they venture afield. More keenly aware of what they do not know, they are more likely to be careful and more inclined to think through a course of action before actually initiating the action. To be sure, as they get more experienced, acting becomes more "instinctual" and less deliberate. After a while, one anticipates how a deer will approach a tree stand in a particular setting, or the direction a pheasant is likely to take once it flushes from the tangled fence row. Beginners, whether youngsters or adults, are almost always slow to react at first, often so slow as to pass up a good opportunity because they needed time to process all of the information required to know if things are right. Adults are less likely to cut their reaction times in advance of full knowledge of what the consequences of acting might be. Call it maturity, or wisdom, or good judgment; adults in general have more of it than teens or twenty-somethings.

With traditional recruitment of youngsters flagging and recruitment of adults increasing, the average age of hunters is necessarily increasing. Insofar as age is an indicator of responsibility and thoughtfulness, this should mean that on the whole the nation's hunters are getting more responsible and that the sportsman's ethic, which may once have been merely a statement of good intentions, is increasingly a description of actual hunter behavior. This said, we still have to face the fact that there are features of the hunt that tempt even the most responsible to let go.

Separate Worlds, Divergent Meanings

The misunderstanding between hunters and nonhunters is not going to disappear. The gulf between the worlds of the hunter and of those who now live on or in proximity to the land that remains huntable is large and growing. Suburbanites and exurbanites see men with guns driving or walking the back roads or slipping furtively through the woods; they hear gunshots around them and assume the worst; and then they see the carcass of a deer draped over a top rack or the head of a deer hanging limply from the back of a pickup truck and cannot help but imagine scenes of violence and bloodshed that yielded this result. Who are these guys? How can I possibly feel at ease with them running around just beyond my property line? Some hunters try to introduce themselves to people whose land abuts the area they hunt in sincere effort to allay fears and misconceptions. Generally, they get a cool reception.

The problem is exacerbated by the fact that the people who hunt are not always the best of ambassadors. Even though most hunters are now city or suburban dwellers themselves, most, as we have seen, have rural roots and most are working or lower middle class. Only eleven of the thirty-seven hunters I interviewed had graduated from college. While this compares favorably to the percentage of college graduates in the population as a whole, it is well below the average educational level of the people who are moving into the countryside to find their own little patch of Eden. Although hunters and the urbanites who have moved to the country might have rubbed shoulders uneventfully in the hustle and bustle of the city, class differences show up almost immediately in exurbia and make some hunters' efforts at communication awkward and even counterproductive.

Even when class differences are nonexistent, the hunter is often at a disadvantage because of the context in which the meeting takes place. I remember coming out of the woods after a morning hunting deer along an old tote road that now divided two recently created house lots. A husband and wife were in their backyard pruning an apple tree in anticipation of the winter snows to come, and I waved cordially to them. The man waved back and then hesitantly walked toward me. I stopped and waited for him to draw near enough to converse. As he approached I offered an innocuous "nice day to be outside" or some such empty greeting. He said, "You know, we really don't like hunters coming so close to the house." He wasn't angry or even alarmed. This would not be the start of a confrontation. I said I understood, but that this path was one of the only points of access to a large tract of land that was ideal for hunting, not only because it was good habitat for game but also because the hunting took place far away from the houses that were strung out along the road frontage. By the time he and his wife saw a hunter, I assured him, the hunter's gun would be unloaded. I was being as calm and professional as I could be. And then I realized how I must have appeared to that couple: I was unshaven and the exertion of the trek out of the woods in an insulated jacket and woolen underwear had left me drenched in perspiration and probably a tad flushed (Had I been drinking? the fellow might well have wondered). I looked disreputable, and had he gotten a whiff of me, he'd no doubt think I'd not bathed for weeks. Under the circumstances of our meeting, even I, a college professor, was unable to reassure him. How on earth would Cal Jones, the carpenter who voluntarily becomes unemployed during the hunting season and lives in the back of his truck for several days running while pursuing deer, have fared conversing with the nervous property owner?

When men and women put on blaze orange hunting vests or camo, they temporarily lose their individuality beneath the layers of symbolism loaded on the image of *hunter*. Anxious landowners do not see a college professor or someone whom they might well have trusted to build their house or fix their car. Nor do they see a woman who teaches kids with special needs or the woman who might have prepared their will. They see a figure about whom they have heard terrible things and whom they reflexively distrust. Guns mix with images of bloodlust and uncontrolled impulse to produce an image that throws otherwise liberal and tolerant folks into fear and loathing.* It doesn't matter in the least that the risk of being shot by a hunter pales in comparison to the risks involved in descending the stairs to one's own basement, never mind the risk entailed in driving to the nearest convenience store for the bottle of milk one forgot to get on the way home. What matters is that hunters appear unbidden and anonymously. When challenged, they seem either surly and defiant, or furtive, as if they have done something of which they are ashamed. Who wants *them* skulking around?

Unfortunately, donning blaze orange or camo doesn't transform the wearer into a saint. Just as the nonhunting public loses the capacity to engage hunters as individuals, hunters, too, lose something in the process. The hunting garb becomes something of a mask. The anonymity conferred by the mask invites the wearer to test the limits of civil society. "Yes, I know that if I infuriate some silly-assed landowner, the reputation of all hunters will suffer, and worse, a blizzard of posted signs might appear in the area by next fall, but, damn it all, I wasn't doing anything dangerous. They just don't understand." Hunters in general have amassed a quite impressive record resisting the temptations of behaving recklessly; temptations that are elevated by anonymity, by the low likelihood of being caught, and by the invitation to free-wheeling that lies at the core of our cultural imagin-

---

*My favorite example of just this sort of thing comes from John Mitchell's wonderfully balanced assessment of hunting (1980). At one point, he is interviewing a suburban Connecticut woman whose expansive property is overrun with deer and has become a magnet for local deer hunters, whose killing ways she deplores. She deplores them to such a degree as to be moved to say, in behalf of her principle of do no harm, "I could just kill them [the hunters]."

ings of the wild. Certainly, this is how hunters think of themselves: they are men and women who understand full well what they could get away with and yet deliberately choose to exercise self-restraint. The disrepute in which they are held by the nonhunting public in general, and by landowners in the suburbs and exurbs in particular, seems to them completely unwarranted and unfair. Worse, it seems hypocritical.

Hunters, to a person, expressed contempt for their critics who deplore killing animals but have no qualms about eating meat bought in the market. Vegetarians don't fare better—"How much wildlife habitat has been destroyed to grow vegetables?" Peter Carbone asked rhetorically. The sense of self-restraint that is at the core of hunters' sense of themselves is thickly interwoven with the pride with which they take at least symbolic responsibility for where their next meal comes from. To be condemned for having (or hoping to have) blood on their hands by people who prefer to have their killing done by others, in places far removed from their comfortable dining rooms, struck everyone in my sample as deplorable. Some hunters shared Zane's angry dismissal of those who posted their land; others seemed almost resigned to being regarded as disreputable. Whether angry or resigned, all the hunters took comfort from and pride in their self-restraint. Whatever others might think, they knew they were trustworthy.

Pages back, I raised the question of whether the sportsman's ethic was merely a fig leaf. I think the matter can now be considered evenhandedly. If the sportsman's ethic is taken to be a literal description of the behavior of every hunter on every outing afield, it is certainly a fig leaf. As I've insisted, there are but a few "bad apples," but there also is a little bit of bad apple in most, if not all, hunters (and, to be fair, in all, or at least most of us generally). Hunters now and then wander from the straight and narrow. Acknowledging this is hard in the face of mounting criticism and the very real prospect of losing the privilege of hunting altogether. But acknowledging lapses is in fact crucial precisely because the infrequent lapses underscore what is most remarkable about contemporary hunting and hunters—hunting is safe and getting safer, and hunters are not simply hiding behind the rhetoric of self-restraint. Self-restraint is deeply embedded in the culture of sport hunting. It is practiced and reinforced, not as an abstract code produced for public consumption, but as a deeply held conviction: self-restraint is what crucially separates hunters from killers.

VII

THE FATE OF HUNTING

For much of our nation's history two contradictory narratives have vied for defining us as a nation and a culture. One is triumphal: the West was won; abundance is ours; the evidence of progress is everywhere apparent. In America everything gets improved. The other is elegiac: we have degraded our natural environment and turned our back on the craftsmanship, self-reliance, and sense of personal responsibility that made us great; in America progress has come at great cost to our natural and cultural heritage. The reason these two opposing narratives persist is simple: they are both true. Our national narrative is a sinuous alternation between triumph and loss. Our greatest accomplishments almost invariably become etched with loss. Dams built to control floods may make floods less frequent but more de-structive when they do occur, and the dams are now known to mess up riverine ecology. Our agriculture is the envy of the world, but our farming practices are wasteful and often ruinous of the land and water. And so it goes—every advance carries with it, it seems, an environmental burden.

Hunters are bearers of this contradictory narrative. Indeed, they have been key players in both versions. As we have seen, hunters have long un-derstood themselves to be the embodiment of the spirit that has made America great. The values reaffirmed by the hunt are the values that un-derpin democracy by sustaining the ideals of independence, self-reliance, and personal responsibility. At the same time, hunters are acutely aware of the toll Americans have taken on the environment. They certainly know, firsthand, about sprawl. In fact, many hunters have a direct stake in it. Rod Granger, the savvy contractor who had built a huge house for his small family tucked well back from the rural road, was not alone in grasping the irony that his livelihood depended upon ruining open land. "I'm slitting my own throat," he observed.

No one with whom I spoke could be said to be an optimist when it came to assessing the future of hunting. To be sure, some felt pretty confident of their own future hunting opportunities. Russ Farina had purchased several parcels of land in the western part of the state, an easy drive from both his home and office, to assure him of a place to hunt. Others, like Rod Granger, banked on trips to Vermont, New Hampshire, and Maine, where open land remains plentiful and the locals are more supportive of hunting. Even many of the less affluent, especially those who, like Fred Jenkins and Zane Zacharias, have close ties to landowners who welcome them onto their land, have been able to make arrangements of one sort or another that they are confident will enable them to keep hunting for as long as they wish. As a result, many hunters were not personally discouraged; but without exception, everyone said he or she felt sorry for the next generation of hunters.

The older hunters were especially rueful. However comfortable they were with their opportunities to enjoy hunting, hunters over the age of forty have seen too many coverts lost to the remorseless expansion of suburbs, shopping malls, and freeways to be sanguine about the future of hunting. Even young hunters like Sioban Osgood, Zane, Ian James, and Jack Wysocki have seen places they had hunted lost to development or rural gentrification. Though they were not morose, none of these young hunters had the optimism one normally associates with youth. Zane was the least perturbed about the future: "They'll always be places to hunt around here," he said, confidently. When I pointed out that hunting areas elsewhere were fast disappearing, and that might mean locals like him would be swamped by hunters coming from other parts of the state, he grew silent for a moment. The prospect was unsettling. He shook his head in reluctant agreement and responded, "I'll find decent places to hunt, somehow." It was as though willing it would make it so.

Though their own experiences with loss were still few, the young hunters had, of course, all listened attentively to their fathers, uncles, and older hunting companions talk of past hunts. Even when the point is not made explicitly, it is hard not to conclude from such talk that the present is but a pale imitation of the past. And yet, when I raised the fact that no one living had been able to hunt wild turkeys in Massachusetts back in the "good old days" (because there were none to hunt), or when I pointed to the fact that the deer herd in Massachusetts has probably never been larger, certainly not in the lifetime of any living soul, the sense of decline and loss was scarcely dented. These were not gloomy or brooding people I was interviewing. Even the people who were enduring a long rough patch, people like Robert Swipe and Paul Julian, struck me as resilient and fundamentally optimistic about life in general. So why were these men and women so pessimistic about the future of hunting?

"We're an endangered species."—Russ Farina

Russ Farina was one of a number of people who wryly observed that hunters were fast becoming an endangered species. This view echoed in almost all of my interviews, arising from one or another of three very different sorts of concerns. For some, the steady loss of land suitable for hunting was the primary problem. Others pointed to overpopulation, which, from their point of view, was driving habitat loss as well as accelerating resource depletion, including wildlife. Still others saw the threat to hunting in cultural and political terms—hunting, they sensed, was becoming objectionable to more and more people. These threats to hunting are, of course, not mutually exclusive. Indeed, several people saw all three threats springing from the same source: an increasingly affluent urban society. Too many people have too much money with which to escape the dangers of the city and the boredom of the suburbs. They move to the countryside with their computers, modems, and fax machines and impose their own values on the land, values that do not include hunting.

The loss of land on which to hunt is a relatively new and rapidly accelerating threat to hunting. Before suburbanization became unchecked sprawl, the major gripe hunters had was the relative scarcity of game.*

---

*These same responses were reported in an interesting study commissioned by the Izaak Walton League of America (1999). The report usefully names the problem a loss of "hunter habitat," public and private lands open to hunting that are also good habitat for game animals and are not overpressured by hunting and other recreational users.

Older hunters remember how hard it was to see, much less kill a deer in Massachusetts. Al Cazminsky remembered hunting on his family's farm as a kid: "I went deer hunting for years without even seeing one, so I gave up deer hunting and hunted for rabbits and pheasants instead." Bill Crafter, the retired schoolteacher, hunted deer in Vermont when he was young because "there just weren't deer in Massachusetts." Wildlife officials confirm these impressions. They estimate that in the 1940s and 1950s the deer herd, mostly concentrated in the western third of the state, numbered less than ten thousand. In the 1990s, the annual *kill* by hunters approached this figure, and in 1997, it topped ten thousand. It is now estimated that the Massachusetts deer population is somewhere around eighty thousand.

Despite the fact that hunters can now expect to at least see deer (getting a decent shot at a buck is another matter entirely), almost all remarked on how they now had to hunt on terms that were less agreeable than when they began hunting. Favorite coverts are now posted or, worse, permanently lost to development. The rapid decline of agriculture and the intensification of farming on the few farms that remain in operation have meant that the New England landscape is growing less and less varied.* While there is plenty of forest, more and more of it is homogeneously mature, which means that it is declining habitat for most game (as well as nongame) species. It may look wild to the casual observer, but "wildness" does not necessarily indicate diversity. Nonhunters may not understand, or if they understand, they may not care, that virtually all game species, birds and mammals alike, require a lot of early successional forest, the kind of growth that takes over an abandoned pasture or that reclaims an area

---

*Farm methods have undergone a radical transformation in the past several decades. The typical New England farm contained woodlots that were continually harvested for firewood and well as saw logs. This meant that there was a mix of maturing and young forest. Topography and technology conspired to add to this mix of woods hedgerows, often ten to thirty feet wide, dividing fields from pastures. These hedgerows, many of them tracing stone walls, grew up in tangles of briars, wild grapes, and scrub that made ideal habitat for small mammals, song birds, and their large cousins, the game animals and predators. Add to these the myriad water courses and wetlands that dotted the New England countryside, and one gets a classic patchwork landscape that was host to a large variety of wildlife species. Nowadays, by contrast, farmers straighten water courses and clear hedgerows right to the property lines, the better to eke out every last bushel of corn or bale of hay. Farm fields now more nearly resemble manicured suburban lawns than the older diversity, and they are home to a much more simplified biotic community.

that has been logged. In a state like Massachusetts, this process is rapidly producing a profoundly ironic situation. In formerly rural areas in the western part of the state, where mature forests are now abundant and much land remains open to hunting, game populations are declining. Forty years ago, the bulk of the state's white-tailed deer were found west of the Connecticut River. Now, there are far more deer in the central and eastern parts of the state, where the bulk of the state's population lives. Coyotes are also now common in the eastern part of the state, as are turkey and geese.* The expanding patchwork of suburbs has created lots of early successional growth which, though fragmented, has been attractive to wildlife. Though wildlife is abundant, this sort of landscape provides precious little land suitable for hunting: it is either forbidden by town ordinances or private land posting, or it is unsafe because of the density of housing.

Hunters report feeling hemmed in. The range of choices of places to hunt grows steadily narrower, and even though the ranks of hunters has been thinning, as I noted in chapter 1, the places that are suitable for hunting are necessarily more pressured. Quiet and uncrowded places to hunt are still available, even in densely settled southern New England, but they are dwindling. With the further stipulation that there be a reasonable chance of finding game, the choices quickly decline. Indeed, most of the men and women with whom I spoke indicated that they hunted in some areas, even though game there was scarce, because they could have the place to themselves. This was especially true for the casual hunters like Rob Collins, the planner. Rob is a good example of the hunter who needs to get out with a gun in the crook of his arm on a few crisp autumn days, but who does not itch to bring home a pheasant or a rabbit. It is now enough for him to approach one of those hedgerows of tangled saplings and vines—a likely spot to find a rabbit, grouse, or pheasant—and tense with anticipation of a flush. Those moments of intense alertness are satisfying in themselves, even if no bird or rabbit materializes. "It makes me feel good to concentrate mentally and physically," Rob explained, "especially after a

---

*This situation has begun to pose real headaches. Geese have taken over parks and reservoirs. Collisions with deer are on the rise (as are the much more dangerous collisions with moose). Bears at backyard bird feeders are fast becoming commonplace. There are now even reports of nuisance turkeys. In Danvers, a suburb of Boston, a flock of turkeys has repeatedly set upon a letter carrier who now refuses to deliver mail in some sections of town out of fear of being attacked by the turkeys.

long week of concentrating only mentally." As long as there are a few such tangles within easy reach, Rob will be personally satisfied—but he also knows that something very important is being lost.

Most of the hunters I interviewed, however, are not as casual as Rob has become about their hunting. It is not that they are filled with blood lust. They want to *see* game. If they get a shot and bring down a bird or bag a deer, so much the better, but seeing game is the important thing. Hunters say, in one way or another, that what they would like are places that are uncrowded and that have good habitat and sufficient game to provide a reasonable test of their savvy. Though many complain of not having enough time to scout out such places, the scouting itself is enjoyable, and, over the years, even if few new coverts are actually discovered, this scouting adds to a hunter's fund of knowledge about nature and game animals. One important reason why scouting is difficult to do is that hunters increasingly have to travel considerable distances to find promising spots. The lucky few who, like Elaine Stebbins, the taxidermist, can walk out their door and within minutes be in good cover know that they enjoy something rare and getting rarer. Elaine chalked this rarity up to one factor: overpopulation. "Too many people chasing too little land," was how she put it. I asked her if she meant that there were too many hunters. She shook her head no, but she went on to note that even though the number of hunters has been going down, there may still be too many hunters for a small and densely populated state like Massachusetts to accommodate. That is why so many of the hunters she knew, including her husband, hunted outside Massachusetts.* Her property abutted state forest land. But she recognized that it was only a matter of time before the state lands around her would become surrounded by homes, and like a tree that has been girdled, the wildness— and wildlife—would die. "I hope I don't live to see it," she said, shaking her head, "but I know it's coming."

With fewer opportunities to hunt nearby, hunters now must plan more and commit more time and, of course, more money to hunting. Planning, traveling out of state, and hunting for new coverts all add up to significant

---

*I know of no comprehensive study of this practice but almost all of the active hunters I interviewed, including several of the casual hunters like Rob Collins, hunted outside Massachusetts. Nearly half had or once had camps in Vermont, New Hampshire, Maine, or New York to which they went each fall. Several traveled more or less annually to Pennsylvania to hunt for deer. A much smaller number, six to be exact, had gone one or more times out West or to Alaska in pursuit of big game. Many more dreamed of doing so one day.

investments of time for the committed hunter—but it is not time spent actually hunting. Several hunters noted the irony involved. Karl Woichek, the teacher who yearns for the day when he will have a chance to go to Alaska or Montana or South Dakota (or all three for that matter), noted that he spends far more time thinking about hunting than he does actually hunting. When he was a kid growing up just outside of Worcester, he could simply change clothes after school, walk a couple of blocks, and be in a field or woods. Even though he was not then the avid hunter he is now, hunting then required virtually no planning or preparation. Open land was abundant and, more important, readily accessible.

It may come as a surprise to people unfamiliar with hunting, but even in as densely settled a state as Massachusetts, easy access to local hunting opportunities was common until quite recently. Twenty years ago, for example, I had my choice of eight coverts, each containing robust populations of grouse and pheasants, that I could reach in less than fifteen minutes from my door. I could easily hunt for two hours and be back home, feed the dogs, shower, and be on campus by nine o'clock (or, if there were no afternoon meetings, I could do the same sort of thing in the afternoon and be home in time to start the evening meal). Only two of those local coverts remain huntable, and each of them, understandably, now absorbs far more pressure from hunters who, like me, have fewer and fewer places in which to hunt. This mounting pressure has negative impacts on both the cover (which gets tramped down) and on the ability of game species to maintain numbers adequate to ensure robust reproduction. Now, even though my work schedule is far more flexible than most people's, almost all of my hunting is confined to weekends because I cannot drive an hour or two to hunt and be able to get back in time for classes, office hours, and committee meetings. My situation, judging from the men and women I interviewed, is typical of most hunters' experience. "The best is behind us," is how Mac Braziel put it. He was talking not so much about the quantity of game as he was reflecting on how hunting was once unencumbered. One just went, and one didn't have to worry much about rubbing elbows with other hunters.

This sense of loss and decline is fueled by another factor with which Mac was also familiar. Mac, as I noted, is the father of two children, the eldest a daughter who was in the tenth grade when I interviewed him. She overheard bits of our conversation and, from time to time, interjected a comment that made plain her disgust for hunting. After one such interjection,

Mac shook his head and shrugged. "It's the schools. They teach kids that crap." Mac wasn't the only parent who had to deal with opposition to hunting at home. Keith Jones's oldest, also a girl, made her rejection of hunting plain. Robert Swipe's daughter, considerably older than either Mac's or Keith's, had become a vegetarian some time ago and now was also an animal rights advocate. She was also adamantly opposed to guns and for a time would not let her young toddler visit his grandparents because there were guns in the house (even though they were all out of sight and securely locked away).

Everyone I interviewed acknowledged that hunting was no longer as accepted as it once was. Only one of the hunters I interviewed, Jim Ramada, reported an instance in which he had a hunt deliberately ruined by anti-hunters. Far more common were reports of looks of reproach or of hostile comments from passersby. Several hunters noted that if they are wearing blaze orange or are walking alongside the road with a gun or bow, the rural etiquette of returning a wave is often not honored by passing motorists or pedestrians. Zane recalled that in the preceding fall, he had stopped by the local convenience store on his way to the check-in station to have the deer he had just shot officially examined and tagged. A woman, relatively new to town, expressed her disgust when she saw the dead deer in the back of his pickup. He dismissed the lady disparagingly but acknowledged that this did not bode well. "I thought I only had to deal with that sort of thing in Amherst [the home of the university he had attended]," he said.

Though the intensity of concern varied considerably among them, all of the hunters acknowledged that support for hunting was eroding and that hunters were increasingly stereotyped as coarse and irresponsible types. Only a handful took seriously the opposition to hunting mounted by animal rights organizations such as PETA. Most, like Zane, dismissed animal rights advocates with the scorn reserved for kooks and cranks.* Whether because of a refusal to reciprocate a friendly wave or nod, or because of a

---

*It is significant that those most concerned were men and women whose children had embraced one or another version of an animal rights opposition to hunting or who feared that they would get "infected" when they entered school. Generally, hunters from the eastern part of the state and those who lived in one or another of the college towns in the western part of the state were more aware of the animal rights movement. Hunters should be more concerned than they are. Animal rights sympathies are clearly becoming more widely shared, and animal activists have grown far more sophisticated in uses of media and the political process, particularly ballot initiatives (Minnis 1997; Muth and Jamison 2000).

direct denunciation, hunters, to a person, knew they were on the defensive. Even if they had had no difficulties themselves with landowners, they knew that hunter-landowner relationships were getting more and more problematic, in part because some hunters abuse the privilege landowners extend, but also because landowners are less and less culturally linked to a sympathy for, or at least an acceptance of, hunting. This sense of defensiveness colored the way hunters talked about recruitment to the ranks of hunters, particularly their feelings about encouraging their children (and in a few cases, grandchildren) to develop an interest in hunting.

Recruiting the Next Generation of Hunters

Hunters are well aware of the fact that hunting is "in trouble." They know that the culture no longer supports hunting unconditionally. A few acknowledged that outright opposition to hunting is now a significant force and a cause for serious concern, though most hunters I interviewed believed that opponents of hunting were few in number and socially marginal. Access to good hunting opportunities was the primary concern. The steady loss of open land suitable for hunting made the parents (and a few grandparents) of young children fret over whether to encourage their children to take up hunting. And some parents, as I have shown, were grappling with the fact that one or more of their children had begun to actively disapprove of hunting. Even hunters who want their children to love hunting as much they do recognize that fewer hunters may not be an altogether bad thing. With suitable habitat likely to continue to decline, what decent habitat there is will surely be degraded if the number of hunters rises appreciably. Even if each and every hunter were a paragon, the sheer press of numbers could ruin things.* Numbers, as well as behavior, matter. A few ignorant or selfish people can make a real mess of things. So can the cumulative effects of hundreds or thousands of considerate people. Hunters feel torn between a desire to encourage more people to share with them

---

*This is exactly what is happening in many of our parks and officially designated wilderness or back country areas. Quite apart from the inevitable "rotten apples" among the ranks even of backpackers or wildlife observers, the steady parade of vibram-soled souls is taking a serious toll on our nation's wilderness areas. Wildlife, too, can be negatively impacted by the press of wildlife watchers. For example, the National Marine Fisheries Service issued a report in 2001 that concluded that the booming whale watching industry "is having potentially adverse effects on the whales" (*Boston Globe,* August 23, 2001, A6)

the experience of hunting that they dearly love and the desire to protect what's left of good hunting from the crush of too many people chasing game on too few acres. Even if game species can be managed so as to ensure robust reproduction, the total experience of hunting would be changed for the worse if more people hunted.

Since most hunters do not see a bright future for hunting, it is easier to shrug off their children's disinterest in the way Mac Braziel did when he said of his son's disinterest, "He won't be missing much anyway." Mac said this with a distinct look of resignation mixed with disappointment on his face. He knew that, in fact, his son would be missing something that Mac felt was important—hence the disappointment. Mac also knew that in all likelihood, even if his son did take up hunting, it would not be the sort of hunting that Mac has experienced—hence the resignation.

Robert Swipe had the same resigned look and the same inflections in his voice when he described the moment he realized that his son had no interest in hunting. They had gone out together to a wooded area known to hold deer that was bounded on three sides by roads that the deer would rather not cross. Their plan was to separate, each moving parallel to one another with a road to the side of each of them, in hopes of moving deer to one another. Robert had, in fact, flushed a deer that seemed headed in the general direction of his son. He stopped and waited with anticipation for the shot that would be his son's first at a deer. There was no shot. Slightly disappointed that his son hadn't had an opportunity, he continued with the plan. When he approached the road at the end of the wood lot, he expected to see his son. He waited for a while, but when his son failed to appear, Robert began to grow a bit concerned. As he tells it, "I swung wide toward the line of travel he would be on and worked my way back toward the car. I called out several times and got no response and really got concerned, though I couldn't figure out how anyone could get lost in such a clearly bounded patch of woods. And then I got to the car, and there he was, asleep in the front seat." He and his son had a heartfelt talk, and his son has not gone hunting since. "Too bad, too; he is a good shot and could have been a good hunter," he said sadly. He now pins his hopes on his young grandson, but, as I've already noted, that hope seems doomed since Robert's daughter hates guns with a passion equaled only by her disgust for hunting.

By contrast, Jim Ramada, the self-employed painting contractor, was as sure as a parent can be that his two young daughters would become avid hunters or at least avid outdoorswomen. Though they are still far too young to hunt or even to be introduced to guns, Jim took them with him for hikes in the off-season. The girls were thrilled to have their dad unlock the mysteries of the woods for them, and the oldest, who had just turned five the week before I interviewed Jim, was, he reported proudly, quite a trouper. She would happily tag along through tough terrain and didn't complain, even if the weather turned nasty. Jim, as I noted earlier, lives in a town that is well-known for its acceptance of diversity and its support for women's equality. When I pointed out that feminism had perhaps made it possible for girls to feel comfortable embracing so archetypically masculine an activity as hunting, he was bemused. Jim was quick to make plain his dislike for "feminism" (which he implicitly associated with lesbianism, animal rights, and opposition to most of the folkways he was attached to), but he and his wife were resolutely committed to the proposition that girls should be able to hunt, play hockey, or become fighter pilots.

There is an irony here that did not escape Jim. He observed that when he was young, almost no father would have dreamed of taking a daughter hunting. Now, it is possible to do so without raising an eyebrow. Indeed, he and his wife were relieved to live in a town where their daughters are encouraged to break all the encrusted gender barriers. At the same time, though, Jim worried that his daughters would be exposed to strong opposition to hunting once they entered the city's school system. The town's gender-bending liberalism carried with it values that Jim and his wife flatly rejected, not least of which was the "Bambi syndrome" that Jim felt had thoroughly infected the town's school system. Jim shook his head as he observed, "If they do take to hunting, they are going to have to put up with a lot of grief from the kids around here."

I wanted to introduce Jim and his wife and daughters to Sioban Osgood. Sioban grew up in the country and attended schools that drew their students from several sparsely settled rural towns, only one of which had what could be called a "center." Sioban recalls being occasionally teased about being a tomboy, but she had easily shrugged off such taunts, if only because she was generally admired for the range of her competency: she got good grades, was a standout swimmer, and was good with a hammer and Skil saw as well as with a shotgun. It was not until she went away to college that she encountered what Jim feared would await his girls: opposition, not to women hunting, but to hunting itself. Sioban's fellow students thought her other "manly" skills (carpentry, roofing, etc.) were "cool," but hunting was definitely uncool. She recalled having many heated discussions about hunting and animal rights in the dorm. Of course, it is one thing for a smart, self-possessed, young woman like Sioban to hold her own in an argument about the morality of killing and eating animals. It is quite another thing for a six- or ten-year-old to resist the pressures of peers and teachers.

Recruitment to hunting, until quite recently, occurred in a cultural context in which fathers taking sons on their first hunt was regarded as a wholesome thing. Boys whose fathers did not hunt, in fact, had a hard time getting a chance to hunt, unless an uncle or a family friend took the lad in tow. Even people who disliked hunting were loath to condemn the father-son bonding on which recruitment to hunting so heavily depended. After all, this was one of the last activities in which sons could follow in their fathers' footsteps. In most other ways, the number of continuities between generations have shrunk, and with the pace of technological change quickening, talk of continuity between generations may soon be entirely the provenance of historians. Mac Braziel has given up: his daughter rejects his values, at least as regards hunting, and his son doesn't even show curiosity about, much less interest in, going hunting. Among the older hunters whose children are adults, only a handful of the men (only the Osgoods and Margaret Jacques had adult daughters who hunted) still hunted. Hunting, once largely a family affair, is much less so nowadays.

Kids go their own ways and develop interests as a result less of parents than of the schools and peers. I also encountered several hunters who were not interested in encouraging their children to hunt. Tommy Worthington's oldest two children were girls, and, unlike Jim Ramada or the Osgoods, he wasn't inclined to encourage them, in part because he wasn't sure it was appropriate (even though he and his wife hunted together before their first child was born), and in part because he was so busy keeping his business in the black that he wanted his hunting time to be just that: his. Whether this attitude will change when Tommy's youngest, a boy, gets old enough to tag along, as Tommy had with his own father and uncles, remains to be seen. Recalling the fact that Tommy remembers his youthful outings with considerable ambivalence may help explain his diffidence about introducing his own children to hunting. Russ Farina was similarly not heavily invested in hunting with his son. He was not making any overtures to him: "If he asks, I'll take him along and help him get started, but I'm not pushing it at all. He's got to decide for himself." So far, his son hasn't shown much interest.

182

Parents are more rushed than ever. Moreover, hunting seasons are short. It takes time to prepare a child, to explain how to handle a gun safely and how to hunt—what kind of cover to look for, how to read signs of animals' presence, how to move safely through dense thickets with gun at the ready. Time spent this way can be deeply rewarding for both parent and child, but it can also be frustrating for one or both. Some parents are clearly not interested in the hassle. In other cases, divorce has made passing on the tradition of hunting harder. Had Keith Jones not had custody of his children, it is doubtful that his sons would have become so eager to hunt. Since few women hunted a generation ago, there are very few single moms in a position to introduce sons or daughters to hunting. Noncustodial fathers face an uphill struggle finding the time to take a child hunting.

Not all hunters learned to hunt at the father's side. The other traditional source of hunter recruitment has been the invitation of a schoolmate or neighbor. Mark West began his bird hunting with friends. His father did not hunt, but his parents did not object to his buying a used shotgun and going out with friends from school and the neighborhood. Robert Swipe never knew his father, and his mother hated guns and refused his requests to be allowed to own one. He began hunting with an older brother, who was less than a model of the sportsman. Ian James, the young man about to leave for a management training course with a nationally franchised fast food corporation, also began hunting with friends from school. But his induction took an unusual twist. One evening Ian came in from hunting pheasants after school with friends, and encountered one of his father's visiting closest friends, who, like Ian's father had never hunted. His father's friend observed Ian coming in with a pheasant and asked Ian to describe the hunt. Struck by the vividness of Ian's account, he grew intrigued and asked if he could accompany Ian on an outing. Ian and his father's friend quickly became inseparable buddies. They now avidly hunt and fish together, each feeding off the excitement and pleasure they have in sharing "firsts": the first deer, the first deer with a bow, and so on. Suspending this friendship was the only cloud over the excitement with which Ian looked forward to launching his career in the world of fast foods. His training and assignment to a franchise operation would clearly require forgoing hunting for at least one season, and he said that neither he nor his friend was pleased by that. But Ian was stoic—it was time to get himself established so that he could marry the young woman who had borne him a son. I asked Ian if he thought he would have become as avid a hunter as he had in the absence of his hunting partner. His response was revealing. "Oh, I loved hunting, but I was not a very good hunter until I began going out with Andrew. When I was with my friends from high school, we'd do stupid things, unsafe things. We knew we were not doing things right because we'd all had hunter-safety courses. We were just wild. When I started going out with Andrew, it was totally different. That's when I really became a serious, committed hunter."

Had Ian gone hunting alone, he probably would have followed the rules he had learned in his hunter-safety course. But when he was with his teenaged friends, the rules were suspended. All it takes is one kid daring the others to do something foolish, and good sense flies out the window. This is as much so in the woods as in our inner cities. Hunter-safety courses, now mandatory in all but one or two states as a condition for obtaining a hunting license, have clearly played a major role in reducing accidents and improving the conduct of hunters, but, as with drivers' education, the power of the lessons conveyed are not equal to the power that arises when adolescence and testosterone mix without the dilution of a more mature presence. Fortunately for Ian, his relationship with Andrew sped up his maturation.

Hunter-safety courses may not innoculate young males from hormonally induced stupidity, but they have made it somewhat harder for young people to take up hunting on their own. Thirty-five years ago, when Mark West began hunting with teenaged friends, he merely needed his parents' assent. With their forbearance and forty or fifty bucks, he acquired his first shotgun and commenced hunting. Today, a young person has to complete an eight- to sixteen-hour course and pass both a written test and a field test to obtain a permit from the local police department to possess a firearm. Only then can he or she purchase a hunting license and legally carry a firearm.* It is possible for young persons to do all this on their own (though they will need a parent's or guardian's signature if they are under eighteen years of age), but with all the distractions now available to young people, only the most determined youngsters would persist to the end. Thus, while it is still possible for a young person to become a hunter without parental encouragement and active support, the more casual or informal avenues of recruitment to hunting, at least for young people, are less

---

*This is the current situation in Massachusetts. The specific requirements vary considerably from state to state, as does the rigor of the hunter-education program.

and less available. This is another reason why the ranks of young hunters are steadily diminishing.

Other recruitment streams, however, have begun to loom larger and involve the entry of adults into the world of hunting. Margaret Jacques was introduced to hunting by her husband. Elaine Stebbins' husband introduced her to hunting after their dating turned serious (he gave her her first shotgun as an engagement present). Several men learned to hunt as adults, encouraged by a coworker or invited on a hunt by a business client. George, the retired engineering executive, came to hunting in this fashion. Like golf, hunting was an extension of his profession, neither purely recreation nor purely work. Again, although numbers are hard to come by, there can be little doubt that with the stream of young recruits to hunting declining and interest among women and, to a lesser extent, adult males in hunting growing, the age and sex structure of new recruits means that the nation's hunters are, on average, better trained, more knowledgeable, and almost certainly less impulsive and reckless than they were a generation or two ago. Hunter-education instructors, as much as they are dismayed by the dwindling numbers of young people enrolling in their courses, uniformly report that both young and adult women are far more receptive than young or adult males to instruction, and while wanting to become accomplished, they are far less interested in counting coup. They don't feel the need to prove themselves by getting a limit of birds the way males, especially young males, do. Adults, especially adult women, it seems, don't need as intensely to prove themselves by shooting more game than the next person. Reduce the proportion of young males in the population of hunters (or drivers or any other population), and the result will be improved safety and a more highly refined ethical standard of behavior (Stange and Oyster 2000).

There is considerable irony in all this. The nation's hunters are better educated than ever. Soon, a large majority of the active hunters will have been exposed to hunter-education programs that go to great lengths to teach safe handling of firearms, game identification, and hunter ethics. And the large majority will also be mature men and women. Though not saints and angels, as I have shown, the typical hunter is now and will be into the foreseeable future significantly more "civilized" than past generations of hunters. Despite this shift, hunters are condemned for their presumed rude impulsiveness and wanton recklessness. As a result, hunters feel more and more cornered, condemned for the sins of their fathers and their own youth and unable to shake the stereotypes with which their critics saddle them. They feel they are being pushed to the edge. To make matters worse, as the culture shifts from beneath them, market forces are also changing the face of hunting in ways that have accelerated the cultural shift away from hunting.

To Market, to Market

Hunters face almost as many challenges on their way to hunting as they face once they are in the field seeking their quarry. An increasing percentage of open land is posted. Suburban sprawl has reduced open lands even further and also requires hunters to invest more time and money getting to places suitable for hunting. And, particularly in densely populated states like Massachusetts, hunting seasons are quite short. The shotgun season for white-tailed deer, for example, runs for two weeks, in late November–early December (archers have an additional three weeks in which to hunt deer). Upland bird hunting, second in popularity to deer hunting, lasts six weeks, from mid-October through Thanksgiving. The season for hunting turkey, a hunt that is rapidly gaining in popularity, is split between a three-week, late spring season (after mating) and a five-day, fall season in early November. No hunting is allowed on Sundays, virtually the last of the blue laws left over from colonial days. Bag limits are also low. Pheasant hunters, for example, are allowed no more than six pheasants per season. Even if seasons were longer and bag limits more generous, work and family responsibilities add their own constrictions on a hunter's time afield. This means that all of the dreaming, practicing with bow or shotgun, and training and keeping a bird dog fit comes to fruition in a concentrated burst of activity for a week or less or, in the case of upland bird hunting, on a weekday here and there and several Saturdays.

For the casual hunter like Rob Collins or Andy Felter, this compression is not a problem. For Mark West and Dick Board, among the most avid of the bird hunters I interviewed, the compression is a problem, eased for Mark because he is self-employed and can juggle his schedule to accommodate morning or afternoon hunts during the week, and for Dick since he recently retired. But for hunters like Karl Woichek, the schoolteacher who has to drive well over an hour from his home to have access to open land suitable for pheasant hunting, the short season and no hunting on Sundays is a huge problem. Since he cannot take vacation time while school is in session, he has, at most, six Saturdays a year in which to hunt with his Brittany spaniel. Hunters who can choose the timing of their vacations typically take time off during the bird or deer season. Many extend their hunting season by traveling to states that have either longer seasons or seasons that start earlier or end later than those in Massachusetts.* A handful, like the self-employed carpenter, Cal Jones, tighten their belts and stop working in the fall.

A moment's reflection reveals the way these factors converge to create a market for a new and rapidly expanding form of hunting—private hunting clubs. Hunters, while fewer in number, are on average better trained and equipped and more highly motivated than their predecessors of a generation and more ago. They are, as I have indicated, also older, better educated, and more affluent than past generations of hunters. Contemporary hunters have invested heavily, both psychologically and materially, in hunting. Bird hunters are especially pinched in this fashion because their investment in hunting is a year-round one; more than that, it is a daily affair: the care and feeding of one or more bird dogs. A dog and hunter who work all year together only to be constrained to bring all this to bear on six pheasants a year are inescapably frustrated.

Private clubs have arisen to relieve this frustration. Private clubs can obtain licenses from the state entitling them to propagate or purchase and release game animals, most commonly pheasant, quail, and mallards, on club property. In most states, such clubs are able to operate year-round (though many limit hunting to the fall and winter months). In Massachusetts, one can hunt on Sundays at these clubs, which for hunters like Karl, who basically have only weekends free, greatly expands their opportunities to hunt. Not surprisingly, Karl has joined one of these clubs, and it has become more than simply a place where he can hunt for the pheasants that are regularly stocked on club lands each weekend. Much of his social life revolves around the club, planning game dinners, competitive dog trials, and shooting events. His club, he was quick to assure me, was not fancy or exclusive. Some years ago, when the area was still largely agricultural, a local farmer gave some land to a group of friends who were organizing a "rod and gun

*To give but one example of this, Vermont is a common destination for Massachusetts' hunters. The deer season there begins in mid-November and ends on the Sunday before the Massachusetts season opens. Upland bird hunting in Vermont begins two weeks earlier and runs for a full month after the Massachusetts season ends.

club." The land formalized the association, and members' dues and an occasional gift or bequest allowed the club to ultimately acquire a couple of hundred acres that accommodate a trout pond and fields, which are managed to provide challenging bird hunting, even though the birds are penraised and released only hours before the hunt.

Some clubs of this sort date back to the early decades of the twentieth century, though most began to assume their current character in the 1960s and 1970s, just when the trends I have been discussing first began to be noticeable. Some clubs, like Karl's, are quite down-home, even plebeian. A few are quite exclusive, and gaining membership in one is akin to joining a fraternity. Memberships and annual dues and assessments in these more restrictive clubs can be quite expensive, easily the equivalent of the cost of membership in an upscale country club, which a few, in fact, resemble. At Karl's club, one would meet plumbers, police officers, small business people, a teacher or two—a slice of the working and middle classes. At the upper end, one would meet doctors, lawyers, and executives—a slice of the upper-middle and upper classes.

In addition to these clubs, whose numbers expanded decades ago but have since stabilized,* a newer arrangement has been growing quite rapidly in recent years: the commercial hunting preserve. Some commercial operations offer seasonal memberships that entitle members to expanded access and other considerations, but members, like the general public, are charged a fixed fee for each of the birds released for each hunt. Less common, but also growing, are the commercial operations offering hunters the chance to hunt large game animals, including wild boar and white-tailed deer as well as exotics like impala and other large ungulates from around the world. Some of these operations amount to little more than shooting fish in a barrel, while others come with no guarantees of success and, largely owing to their expansive acreage, can claim to approximate "real hunting" (Bilger 2001). But whether we are talking about the unpretentious club to which Karl belongs or about a commercial preserve where it costs thousands of dollars to shoot an exotic animal, the question of "real hunting" looms large.

As I have shown, definitions of the hunt have evolved over time. The fail-

---

*Land costs began to soar in the 1960s as the suburbs began their steady march into formerly rural America. Land-use restrictions followed so that a new club faces daunting start-up costs, if they can even obtain the necessary permits and easements allowing shooting, wetlands modifications, and the like.

ure of nonhunters to fully appreciate this evolution is part of what makes contemporary hunters feel as though they are misunderstood. Someone who goes into the woods out of season and blazes away is, from a hunter's point of view, not a hunter, though that person may, in a very narrow sense, be "hunting." To be a hunter is, nowadays, to accept self-imposed "rules of engagement" that keep the playing field level so that the hunter is engaged in a *fair chase* with his quarry.

Like most evolutionary changes, change is rarely even. The present is always some combination of the new and the old. Indeed, part of what drives change is precisely the tension between the newer and the older elements. This is as much true in the physical as in the social world. It is certainly true for the world of hunting. Hunters intensely debate whether it is ethical to shoot over bait, whether it is ethical to use infrared or other night-vision enhancement technologies, and whether "canned hunts" are, in fact, "hunts."* There are, in fact, unresolved debates that regularly appear in the hunting press and on Internet discussion groups aimed at hunters about what constitutes a "canned hunt." Everyone would agree that shooting an animal that is confined in an enclosure so small or so barren of cover that the animal has no means of escape or concealment is "canned" and does not qualify as hunting—shooting, yes, but definitely not hunting. But what about the pheasants that someone at Karl's club puts out in the club's field a half hour or so before Karl and his dog set out? Does the fact that those pen-raised birds can hide in the hedgerows and tangles that club members have created to make it possible for pheasants to elude Karl's dog make his hunting "real hunting?" After all, not only can the pheasants hide, they can also fly off unscathed. On the one hand, many would say that Karl is really hunting because the outcome is by no means certain. The chase is fair because Karl has not manipulated the odds to his advantage. The knowledge he has that a hunter in the wild would not have is that there are birds in the particular area in which he is hunting.

On the other hand, unlike the hunter in search of wild birds, Karl does not

---

*Shooting over bait is an old practice that once was common in all sorts of hunting, from waterfowl to deer and bear hunting. In a number of midwestern states, it is still perfectly legal for deer hunters to put out piles of sugar beets or carrots to attract deer to the place where the hunter waits in his blind. Bags of donuts and such are also, where legal, used to attract bear. In some states, baiting one species is illegal while baiting another is legal. In other states, all baiting is prohibited. Whether legal or not, hunters' regard for such practices is sharply divided. Many hunters would not dream of baiting, even when it is legal in their state.

need to do any of the hard and challenging work of searching out pheasant coverts or learning the ways in which pheasants alter their behavior and their preferred habitat as the fall melds into winter. At the appointed hour in a designated area, Karl's birds await him.* Having hunted pheasants under just such circumstances, I know there is a difference, though I think the difference doesn't speak to fair chase in the same way it would if I were talking about hunting animals in a compound from which there was no escape and in which there is nothing remotely approaching the animal's native habitat. The kind of hunting Karl's occupational and geographical constraints impose upon him is less a test of him as a hunter than it is a test of his dog and how well he has trained him. Karl's only role, beyond that of working with the dog, is that of a shooter. Everything else is up to the dog. Hunting in areas where the hunter knows birds have been released only hours (and sometimes only minutes) before the hunt begins more nearly resembles a field trial in which dogs are put through their paces in competition with other dogs than it does a hunt for wild birds in naturally occurring coverts.

There is another aspect to this sort of hunting that raises some troubling questions about an aspect of the ethic of fair chase. It is worth remembering that fair chase is not simply about keeping the playing field level, though that is certainly a central aspect of the ethic. Fair chase also demands that the hunter respect his quarry. Indeed, this condition is foundational to the ethic itself: the level playing field is desirable because the hunter honors the quarry. To take unfair advantage or to stack the deck is to betray a respect for the creature. It is, plainly put, harder to confer respect of the sort demanded by the ethic of fair chase on pen-raised or captive animals, even if the hunt for them is otherwise a fair contest. Hunters who have a choice, people like Mark West or Elaine Stebbins, will elect to hunt wild birds when they can. They may also hunt on state-owned "management areas," but they freely acknowledge that it is not the same. As for hunting at clubs like Karl's, hunters who have ample opportunity to hunt wild birds during the public hunting season will hunt at clubs or commercial preserves only in the off-season. These hunts are more social—a couple of friends taking an afternoon off to-

---

*I say "Karl's birds" because, in a sense, he owns them: he and his fellow club members bought the pheasants from a commercial game farm as chicks and proceeded to raise them over the summer. Some clubs charge members, in addition to dues, a set fee per bird released. In the commercial clubs, members and guests pay a fee per bird released (whether or not the purchased birds are shot). A few operations charge only for the birds shot, but understandably, their fees are higher.

gether—and are frankly acknowledged as being "for the dogs," that is, the point is more to keep the dogs on their toes than it is for the hunters to hunt.

The difference is that as the hunt becomes more contrived, attitudes toward the bird (or any other game animal) shifts subtly. In the state-run management areas, as Karen DeFazio ruefully noted, hunter behavior sinks to an unbecoming low. Everyone knows that the birds have just been stocked, and since there is no limit on the number of hunters, there invariably are many more hunters than there are birds. As a result, people are drawn into competing with one another (arguments over whose shot was the killing shot are not at all uncommon), which shifts the focus from the bird to "getting mine before the other guy does." When the scene changes from stocking in public areas to stocking in private clubs or commercial preserves, to venues where the hunter directly buys the game to be hunted, this shift is even more pronounced. Even though the element of competition is absent because each group of hunters is assigned an area to themselves, at twenty or more dollars a bird it is hard to honor the wildness of the pheasant, even for those for whom the money itself is not a big deal. Hunters realize that they are paying for the experience, not for the birds, but it is still very hard not to regard those birds as "mine." The normal suspense of the hunt—will I be adept enough to find game?—is replaced by a different and muted suspense: will I get all the birds I paid for?

Either way, whether on public or private lands, the game becomes less a wondrous gift of nature, a gift to be held in awe, and more a commodity. To be sure, this is a very different sort of commodity than was the case in the days of market hunting. But the difference is not so great as to leave hunters entirely comfortable. When hunters worry about the future of hunting, part of their worry is that because land open to hunting is declining, more and more hunting is going to take place on public areas specifically designated for hunting or on private lands run as clubs, commercial hunting preserves, or lease hunting (where one or more individuals pays a landowner for the exclusive right to hunt in a particular area).* As with so many other recreations, this will mean crowded and unpleasant hunting for those who cannot afford

---

*Leases are common in many parts of the country. In fact, many farmers in the South, Southwest and Midwest supplement their incomes from farming with the fees they charge for hunting access. Leases for goose and duck blinds, pheasant and quail hunting, or white-tailed deer can exceed several thousands of dollars for a week, and much more if access for a whole season is involved.

what will almost certainly be increasingly costly private arrangements. Lost will be all but a glimmer of the democratic tradition of hunting, a tradition that gave the ethic of fair chase its distinctive American flavor.

The market has responded to the pinch of busy schedules, to short seasons, and to mounting restrictions on access to land on which to hunt in another way that threatens to undermine the ethic of fair chase. Karl Woichek's dream of hunting in South Dakota or Montana, should he ever realize it, will almost necessarily involve hiring a guide. After all, how would someone utterly new to the area, with only a few days to spend, begin to locate land suitable for hunting, find out who the owner(s) of the land are, and secure their permission to hunt? The only sensible thing is to book the outing with a guide/outfitter who knows the area and has secured the necessary permissions with local landowners. Guides stay in business because the "sports" who hire them have a good time. As I have already made plain, most hunters are not obsessed with bringing game home each time they hunt. They are content and say they have had a good time as long as they see game. However, when the hunter has invested a significant amount of cash for the services of an expert, coming home empty-handed can be an unpleasant prospect. Even if the hunter does not mind, the guide cannot help but worry about his reputation and his bottom line. If too many hunters leave empty-handed, he is out of business.

The guide's uncertainty leads to all sorts of temptations and shortcuts to ensure that clients leave satisfied. The ads for guides in the outdoor publications routinely boast of prodigiously successful hunts. Pictures of smiling hunters standing either alongside several deer or elk carcasses hanging from the side of a lodge or next to the lowered tailgate of a pickup obscured beneath a pile of pheasants convey the promise: "Book your trip with me, and the fellow in the picture will be you." Of course, hunting isn't hunting without the prospect of coming home empty-handed. If there is certainty of success, something's missing. Hunting is reduced to shooting and game becomes simply a target. The pressure on guides to produce "targets" makes it tempting for them to stack the deck in favor of the client. The guide's and the hunter's grip on the ethic of fair chase is inexorably loosened. Sometimes, the ethic vanishes completely.*

*In 1991, *National Geographic* published a long story (Poten 1991) on the exploitation of wildlife by unscrupulous outfitters and clients eager for the hunt of a lifetime. The article also sheds light on the burgeoning traffic in wildlife whose various organs, glands, and sundry body parts fetch lots of money in East and South Asia.

The conditions required for the full embrace of fair chase are being stripped away in favor of conditions that permit, even encourage, tilting the advantage decisively toward the hunter who presents him- or herself as a paying customer. Sporting-goods manufacturers and retailers add their own impetus to this tilt. The months leading up to each new season are filled with announcements of new products purporting to make the hunter a better tracker and a better shot. Vehicles are refined to carry hunters farther and farther into back country in ease and comfort, reducing or removing altogether the combination of physical endurance and woodcraft that hunting ordinarily involves. Some states are moving in the direction of banning electronically assisted sights that were initially invented for military use by snipers and sharpshooters—lasers, assisted night-vision devices, and the like. But the pressures are relentless—there are profits to be made with each new gadget or refinement of last year's gadget. And, like the guides' promotional materials, the meaning of the hunt is implicitly or explicitly defined as "bringing home game." Yes, lip service is paid to the idea that "it is good just to get out—it's the experience of the hunt, not a full game bag, that matters." Still, this rhetoric wears thin when, for more and more hunters, hunting is compressed into one or two outings for which considerable amounts of money have been shelled out.

The critics of hunting will more easily criticize to the extent that hunting becomes less and less a direct, evenly matched contest between hunter and hunted. The more contrivance, the more elaborate the apparatus and the paraphernalia involved, the less hunting can be represented as satisfying either ancient callings or modern ethics. And crucially, the veneration in which game animals are held is diminished. I have frequently heard hunters refer to stocked pheasants as "chickens." I have never heard a hunter call a wild pheasant or a grouse a "chicken." The standing of game animals is lessened by the creep of commercialization because it puts too much emphasis on the kill, on linking a successful outing with game in the ice chest.

The standing of game animals is also being jeopardized from another angle that is not market-driven. As I have shown, some species of wildlife, most notably white-tailed deer and Canada geese, but also increasingly beaver, black bear, and even wild turkey, are fast becoming "nuisance animals," especially in suburbs in which hunting has long been banned. While many communities are adopting a wait-and-see stance, apparently as uncomfortable with hunting as they are with the nuisance animals, many communities are turning to hunters to help them reduce the population of problem animals.* Whatever one might think of this as a "public service," it is not hunting in any meaningful sense of the word. The animals, while not exactly tame, are far from the wily and evasive woodland creatures with whom hunters have matched wits for centuries. Shooting suburban deer or geese is hardly different from the canned hunts that raise so many questions. To the extent that hunters allow themselves to get drawn into "pest control," they run the very real risk of undermining the sporting ethic (Dizard 1999; Dizard and Muth 2001). In the bargain, this service will reinforce the public suspicion that hunters, after all, are really only interested in killing things. If there's a dirty and unpleasant task involving wild animals, let those half-formed humans take care of it. People may be relieved, but gratitude for solving someone else's problem does not necessarily translate into respect. We appreciate the work of grave diggers, but that does not make them esteemed members of our community.

Most hunters simply hunt. They do not pursue exotic game on game farms, belong to private hunting clubs, or hire the services of hunting outfitters. There is little contrivance involved in everyday hunting. The guns may be brand new, but they are based on technology that has been in use in manufacturing firearms for over a hundred years. Indeed, some hunters hunt with guns that were used a hundred years ago. The typical hunter pursues deer or game birds that are indigenous or, like the pheasant, have been resident in North America for over a century. Still, most hunters at least understand why someone might try to gain advantage with a new gadget or a bow made of space-age materials and pulleys that make accuracy easier to achieve. They also can understand why fellow hunters join clubs so that their hunting season can be extended and their dogs will be able to work many more birds than would be possible in the six-week, public bird-hunting season. And most support augmenting native game bird populations with birds propagated for release during the hunting season, as Massachusetts does with pheasants.

---

*There is an unsettled question in all this that is worth keeping in mind. Some wildlife officials and hunters argue that you actually do not have to kill all that many geese or deer or bear to resolve the population problem. The number of animals is only part of the equation. The other part is that these animals have lost their natural fear of humans. As long as no one is shooting at them, there is absolutely no reason not to graze ornamental plantings, take over playgrounds and golf courses, or feast off backyard bird feeders and garbage cans.

Taken together, hunters find themselves with a flank more and more exposed to the barbs and arrows of the critics of hunting. Each departure from the ethic of fair chase and the more encompassing sporting ethic, however small and however reasonable the departure might be (it is not unreasonable for Karl Woichek to join a club that releases pheasants for him and his dog to hunt), adds up to making hunting a shakier enterprise. In fact, in recent years several hunters have written sharp condemnations of contrived game farm or canned hunts, among other departures from the standards of fair chase (Kerasote 1993, 1997; Peterson 2000). As compelling as their arguments are, the problem is that the conditions that make traditional hunting possible are fast disappearing. Hunters are by no means prepared for this mounting criticism. They have, as I have shown, long thought of themselves as embodying ideals that are at the heart of the American national narrative. Men and women just like themselves were the ones who made America great. As Herman (2001) has compellingly shown, hunting became a key element in defining what it meant to be American—it was part of the sense of American ethnicity. In the face of mounting criticism, hunters have become more and more defensive. Long a statistical minority, hunters are fast becoming a cultural minority group. This new status is not gratifying.

From American Ethnic to Minority Group

Virtually every hunter I interviewed spoke defensively when our conversation turned to the general public's perception of hunters and hunting. Almost everyone complained about the way the media, especially the major television networks, portrayed hunting and hunters. The movie *Bambi* was frequently mentioned as casting hunters in a particularly nasty light. Many saw the negative portrayal of hunters as part of a larger agenda that included gun control and restrictions on land uses that they felt were elitist. So, for example, while almost no one favored unregulated use of all-terrain vehicles (ATVs) and many said they hated the contraptions (objections were to their noise and to the damage they did, both on and off the trail), they nevertheless saw attempts to regulate ATVs as an assault on ordinary peoples' recreation. They resented being stereotyped as much as they resented the stereotype itself—the beer-swilling, pot-bellied, gun-happy bozo. As Fred Jenkins put it, "they see blaze orange and think the worst."

In the same vein, hunters felt trapped on the issue of gun control. As we have already noted, nationwide polls reveal that while hunters are less enthusiastic about gun control than are nonhunters, a majority of hunters in fact favor tighter restrictions on access to guns. Many of the hunters I interviewed had no interest in owning a handgun, and many had no objection to banning or sharply restricting ownership of military weapons whose only purpose is killing humans. In fact, only two of the hunters I interviewed were "gun people," more interested in shooting sports than hunting. Four others shot handguns competitively at local gun clubs but also took their hunting seriously. The vast majority of my interviewees were interested in guns only as they related directly and practically to their hunting. No one thought their hunting weapons would be banned, so in that sense, gun control was more a symbolic than a practical concern. Like those who attack hunting, advocates of gun control seem bent on discrediting the traditions and values with which most hunters closely identify. Thus, Bill Crafter, the retired teacher and woodcarver, who acts more like Mr. Rogers than Charlton Heston, was worried about gun control, not because he needed a gun, but because he felt that the freedom to own a gun was a fundamental and defining freedom without which our democracy would be hollow.

Many were also irritated by what they saw as a steadily growing mountain of rules and regulations governing guns and hunting, as well as much of daily life. Even if they agree with the intent of the regulations, it was clear from a number of comments, sometimes about a gun law, sometimes about the motorcycle-helmet law (Massachusetts has one; Connecticut, Rhode Island, and New Hampshire do not), that regulations signal a lack of trust between government and the people. As regulations pile up, people lose self-reliance and the appetite for freedom, the very virtues hunters see embedded in hunting. There was also a cynical aspect to the dislike of regulations. Many saw the government as blundering and inept. With respect specifically to gun control, many argued that there were enough laws regulating guns on the books already and that if existing laws were enforced vigorously, there would be no need for additional regulations.

Hunters were also worried about environmentalists. Protecting land from development was not the problem, though many of the people I interviewed owed their livelihoods, at least in part, to the jobs that arise out of building roads, subdivisions, and shopping malls, a conflict many felt quite intensely. The problem was that many environmental groups want to protect land, not only by prohibiting development, but also by declaring land off-limits to hunting, motorized vehicles, and motor boats. Opposition to logging is also a problem for hunters. As I have shown, virtually all game species (and many nongame species as well), require "edge," places that are in the early stages of recovering from disturbance, whether from agriculture, fire or storms, or logging. Native Americans burned to create edge in order to produce better habitat for the animals on which they depended. Hunters know from practical experience that game is more plentiful in areas where there is a lot of edge. Large unbroken tracts of mature forest may seem wild, but from a hunter's perspective, wild and wildlife are by no means the same thing. As long as environmentalists promote policies that restrict access or make blanket prohibitions on activities like logging that produce edge, hunters recognize that their interests are being ignored.

More than self-interest is involved, though. Hunters feel that many environmentalists disparage them and the way of life they identify with. Hunters, after all, love nature, too, and they are rightfully proud of the collective record hunters have as promoters of wildlife and land conservation. Though most hunters know only bits and pieces of the history of conservation and wildlife management, they all are well aware of the fact that their license fees and taxes on their guns and equipment have played a major role in protecting the environment. To suddenly find themselves cast as anti-environmental is both bewildering and infuriating.

In all of these ways, hunters feel "on the outs." Their love of nature and wild things is ridiculed by animal advocates and many environmentalists. Because they are intimately involved with firearms, hunters feel stigmatized as "gun nuts" or prone to violence by those who advocate more restrictive gun laws. I probed to see if there was any sign of defiance or belligerence, either a posture embodied in Charlton Heston's statement, "You can have my gun when you pry it from my cold, dead hands," or sympathy for militia-type groups. I found none of either. On the contrary, the mood or tone was more one of resignation and sadness, often mixed with defensiveness. "They (those who would like to see an end to hunting) just don't understand us" was a common refrain in the interviews.

This defensiveness arises from a general and broadly shared sense that the culture has lost its direct links to the land. Like the family farmer who has all but completely vanished from the scene, hunters see themselves as one of few remaining groups in our society who are participants in nature rather than spectators. This link to the land and its bounty carries with it,

197

as I showed earlier, a sense of humility and modesty that hunters, whatever the size of their pocketbook and whatever their political allegiances, regard as a necessary corrective to the larger culture's materialism and egoism. Defensiveness also springs from more mundane sources.

Most hunters have had no personal contact with animal-rights or anti-hunting activists. For the most part, the threats to hunting remain, abstract, that is to say, more a matter of *zeitgeist* than of actual displacement. But every hunter has had direct experience with Posted or No Trespassing signs that declare land off-limits to hunters. By all measures, these signs are proliferating and, as much as anything, represent the cultural shift away from an acceptance of hunting.* In some areas, a person can drive for several miles at a stretch along dirt roads and see land on both sides of the road posted, even though there are no houses or livestock or any other signs of human activity that would not mix well with hunting. Hunters sense that they are being hemmed in. They also feel that they are being betrayed by politicians who have lost touch with "ordinary folks."

Most of the men and women I interviewed considered themselves "moderates," and of those over forty, most voted for Democratic Party candidates routinely. Of course, the Republican Party has long been weak in Massachusetts, so voting a straight Democratic ticket is hardly unusual. Still, hunters are less and less inclined to reflexively pull the Democratic lever. In their eyes, the Democratic Party has embraced precisely those things that disparage hunters: gun control, an environmental philosophy that precludes hunting, and even sympathy with animal rights. Although

*There are dramatic regional variations in regards to public access to private lands. In New England, the time-honored practice is that all lands that are not posted are presumptively open to hunting (assuming that distances from roads and buildings meet the legal requirements). In much of the rest of the country, hunters are required to gain the verbal or written permission of the landowner. Groups opposed to hunting have begun to try to get laws passed in New England requiring landowner permission. Although innocent on its face, this change would pose a daunting set of tasks for most hunters: landowners are far from easy to locate; property lines are far from unambiguous; and—a reflection of the length of time New England has been settled—private holdings are frequently quite small and holdings are not necessarily contiguous. To take but one example I know of personally, one rather small area in which I hunt—an area small enough to be hunted thoroughly in two hours or less—involves land owned by seven different owners, three of whom are absentee and, so far as I know, have not set foot on their property in twenty years. It would not be a simple matter to track down every owner, and failing to get all owners' permission, I would be unable to hunt the whole area because it would be almost impossible to avoid the parcels for which permission had not been obtained.

they still vote for Democrats in local elections, statewide and presidential races are another matter. In recent elections, Democratic gubernatorial candidates have run poorly in the more rural parts of the state where hunters are most numerous, even though Democratic candidates for the state house and senate generally win.

In effect, hunters are beginning to see themselves as a "minority group" whose way of life is threatened and whose values are under attack. Nothing in their experience has prepared them for this shift. Indeed, they believe they are bearers of traditional American virtues who are suddenly finding the rug pulled out from under them. There can be little doubt that this sense of minority status is pushing hunters toward conservative candidates in both the Democratic and Republican parties, but especially toward those in the Republican Party. The irony is that while the Republican Party gives lip service to the values hunters hold dear, they pursue policies that favor unregulated exploitation of natural resources. The self-styled "sportsman's caucus" in the U.S. House and Senate earns, as a group, appallingly low ratings on virtually all environmental scorecards. There is not much hunters can do to get themselves out of this trap. They are, after all, a small group. Like other minority groups, hunters will have to learn to choose their allies carefully. It will do no good in the long run to have supported leaders who rhetorically support hunting while promoting policies that are ruinous to the environment. The result will almost surely mean the demise of the democratic tradition of sport hunting. In its place, we will have game farms and hunting clubs and precious little opportunity for public hunting. Clearly, hunters have much about which to worry. The question remains: Why should non-hunters care? What difference would it make if, through attrition or a sharp shift in public law, hunting, like public flogging, becomes a thing of the past? What, if anything, is at stake?

Mortal Stakes

Defenders of hunting typically base their support for hunting on two sorts of claims. The first claim is that without hunting, wildlife management would suffer, and that, in turn, would be bad for wildlife. The second claim is that hunting is traditional, with roots back deep in prehistory; indeed, some claim, the tradition arises at least as much from the evolutionary forces that shaped our biological functioning as from the accretions of culture that have embellished the hunt with a thick tapestry of lore (Shepard

199

1973). While each of these defenses has merit, neither can suffice to justify hunting today. However passionately one might feel about a custom or tradition, arguing "we have always done it" is not persuasive to those who find the activity objectionable, whatever the source of the objection. It is equally clear that modern sport hunters are not, by virtue of their hunting, assets to wildlife management—money raised by the tax on hunting equipment and from the sale of hunting licenses has been utterly crucial to the largely successful efforts to manage wildlife, but the funds derived from hunters could be found from other sources.*

Hunting is at best a blunt wildlife management tool. To be sure, hunting has contributed to keeping some wildlife species in check, most notably white-tailed deer. Where these efforts have been successful, and there are many success stories, hunters have had to be encouraged to suspend, or at least to qualify, the ethic of fair chase in order to maximize the kill (Dizard 1999). The ethos of fair chase, coupled with centuries of lore, strongly bias sport hunters toward seeking male deer with impressive antlers. The management of white-tailed deer, by contrast, requires controlling the number of females.** Were hunters primarily motivated by the old "fill-the-pot" mentality, their goals might more neatly mesh with wildlife management goals, but as I have shown, few hunters nowadays hunt mainly for the pot. It is also true, though, that hunters want to see game: it is in their self-

---

*I do not mean to suggest that finding alternative funding for wildlife management, research, and habitat restoration that now depend heavily on revenues generated by hunting will be easy. I will cite but one example: through much of the 1990s, a coalition of environmental groups working under the banner "Teaming With Wildlife" (TWW) sought legislation that would tax a wide range of outdoor equipment, from binoculars to sleeping bags and hiking boots, in the same fashion as fishing and hunting gear has been taxed since the 1930s. The initiative went nowhere, blocked by the "no new taxes" mentality prevailing in Congress and the executive branch, regardless of which party was in charge, and by the firm opposition of the manufacturers of outdoor equipment. TWW has been succeeded by new legislation which, despite broad support in Congress from both parties, has not been signed into law. The bill, the Conservation and Restoration Act (CARA), if it passes will funnel almost a billion dollars a year into various programs to enhance the environment and prospects for the nation's wildlife. Though all hunting organizations of note strongly support CARA, as they did TWW, this support comes with the recognition that hunting will lose its premier position as *the* major source of wildlife funding.

**Bucks will impregnate a number of females each fall. If the does are protected, as they were from the 1930s until recent decades, the result, all things being equal, will be a steady increase in the herd. This was the principal means of rebuilding the nation's white-tailed deer herds. Similar sex-specific strategies have been a key component in the management of migratory waterfowl, pheasants, and turkeys.

interest to act in ways that ensure robust reproduction so that next year's hunt will be at least as good as last year's.

Hunters have been willing to support the management efforts of state and federal agencies precisely because most of those efforts have been directed at creating and sustaining healthy populations of game species. The problem is that the populations of some of these species have exploded, and as with deer and nonmigratory geese, it is now unlikely that hunting of the sort that has long been practiced can do more than help resolve very localized problems. The dilemma is revealed starkly with the case of the snow goose. For reasons I need not detail, the population of snow geese (a smaller cousin to the more familiar Canada goose) began to explode in the 1980s and 1990s. By the late 1990s, it was clear that the snow goose was destroying not only its own habitat in the fragile environs of its breeding grounds in the far north but also the habitat of many other birds, small mammals, and other creatures. The U.S. Fish and Wildlife Service (USF&WS), in conjunction with its Canadian counterpart agency, essentially declared an all-out war on the snow goose. Virtually all of the regulations that had long governed migratory waterfowling were suspended— bag limits were removed, restrictions on electronic or amplified calls were relaxed, seasons were expanded. The message was clear: snow geese had become a scourge, and their numbers had to be drastically and quickly reduced for their own good.

The hunting press response was less than enthusiastic. The biology of the USF&WS was not challenged, but there was open resistance to endorsing practices that had long been condemned as unethical and unsportsmanlike. George Reiger, a columnist for *Field and Stream* and an often acerbic critic of the USF&WS, was particularly vigorous in questioning the wisdom of encouraging the nation's waterfowl hunters to enlist in what would amount to a slaughter (Reiger, June, 1999, 28–30). As it turns out, the new regulations were challenged by animal rights organizations who insisted that the USF&WS had not carried out the appropriate environmental impact studies. Even if the changes had gone into effect, it is by no means clear that hunters would have lined up in sufficient numbers and with sufficient enthusiasm to achieve the magnitude of reduction that the USF&WS desired. Strange as it may seem to the critics of hunting, hunters are not all that interested in unrestrained killing of game.

Hunting ethics emerged in response to two conditions: depleted game populations as a result of habitat loss and market hunting; and a quite dreadful safety record. Today's hunters, according to what I heard in my interviews as well as from the statistics collected by state and federal agencies, are well schooled in this ethic. Though this does not preclude occasional lapses in personal conduct, on the whole it means that hunters, much as they want to be successful, are not entirely comfortable being thrust into the role of becoming active wildlife managers when it comes to killing on a large scale. Indeed, most hunters would take more kindly to a reduction in bag limits, if the goal was to help a species maintain itself or recover, than to pleas to enlarge the kill. Strange though it may seem, hunters are more supportive of wildlife management when that involves managing species whose future is clouded by declining numbers than they are of management aimed at reducing the numbers animals who have become too numerous.

Hunting has been good for wildlife, both directly and indirectly, not because individual hunters consciously harness themselves to broad wildlife management goals, but because hunters have been effective lobbyists for the critters they love to hunt. Many hunters also have an environmental sensibility that extends well beyond their narrow self-interest, but that is a separate matter. What counts is that the men and women who love to hunt grouse or deer or elk or whatever have been willing to pay for research and habitat enhancement and to accept restrictions on their own hunting in order to ensure that game animals prosper.

Though critics like Ron Baker find this contemptibly self-serving, the fact is that but for this willingness, wildlife in North America would be in far worse shape than it now is. Moreover, though hunters are largely interested in enhancing game, the same habitats that favor game animals are a boon to all sorts of other nongame species. Managing for game does not, as Baker and others claim, come at the expense of other species. On the contrary, hunter-supported wildlife management has been a win-win situation for wildlife in general. As important as these efforts have been, however, it is still the case that nonhunters, who vastly outnumber hunters, could take up the cause of wildlife were hunting to vanish. There is, though, little in the historical record to suggest that the general public will ever become as motivated as hunters have been to support wildlife management programs. Still, other sources of money could no doubt be found, though these funding sources come with a very large liability: they will almost certainly be public funds appropriated by state and federal legislatures. This will inevitably mean that wildlife will be made to compete with all sorts of other legislative priorities. When coffers are full, wildlife may well prosper; but when things get tight, it is hard to imagine wildlife holding its own against national defense or desperately needed social programs. The agencies that depend upon hunters' license fees and the tax levied on hunting and fishing equipment have been insulated to a large degree from the shifting sands of political partisanship and the ups and downs of legislative wrangling over budgets. If the agencies and organizations hunters have historically supported were to disappear, there can be little doubt that wild critters would lose one of their staunchest allies.

Absent modern sport hunting, wildlife could still be managed. Those game species whose numbers pose problems could be controlled by a variety of lethal and nonlethal means, including sharpshooters, nest disturbance, contraception, and trapping. Some present-day sport hunters may refashion themselves "animal-control agents" and offer their services to communities trying to deal with an overabundance of deer, geese, or bear. The general public would hardly notice the difference, except perhaps that they would now see some of their tax dollars diverted to pay for the services of these animal-control agents. Even though the change may go unnoticed by the general public, the change would nonetheless be consequential, if hard to measure with any precision. It's not so much that a tradition would die, though that is a part of it. More important would be the loss of something that is embedded in the tradition of hunting—the sense of immediate engagement with and participation in nature.

Many observers have remarked on how we have been steadily transforming ourselves into spectators. There is nothing wrong with spectatorship per se. In fact, it is plain thrilling to witness highly skilled athletes compete, just as it is inspiring to watch a ballet performance or listen to an orchestra perform a Mozart symphony. Indeed, watching can capture our imaginations and inspire us to improve and enlarge our own skills, even if we remain distinctly inept in comparison to those who inspire us. To the extent that watching moves us to become engaged and active, it is a good thing. But when spectatorship becomes a substitute for active engagement, when watching is our only form of engagement with the "real world," something quite important gets lost.

There are precious few ways left in which we can engage nature as participants. The products of nature upon which we depend for our food, clothing, and shelter come to us in highly mediated forms, so mediated that it is possible for large numbers of our young people to have no way of connecting the wood that frames their dwelling to a forest, or the shrink-wrapped chicken tenders they eat to a real, live chicken. We are losing any sense of where things come from, and if the promises of computer technology are any guide, it will not be long before the real world will pale in comparison to the virtual representations of it we will be able to bring into our homes—all neatly sanitized and packaged: no dirty hands. However realistic our technology may make such encounters with the real world, they can never be a full substitute for the real thing—the exertion, the sweat, the aching muscles, and, yes, the dirty hands. In this sense, it matters little whether we are speaking of gardening or hunting. There is something different, something special about the direct engagement with and the unmediated participation in nature that is involved in gardening and hunting (we can include fishing, trapping, cutting firewood, as well as back country hiking, mountain climbing, and whitewater kayaking) that is utterly absent in wildlife watching or eco-tourism, to name two forms of spectatorship involving nature whose popularity is rapidly increasing.

Direct engagement—dirty hands—teaches lessons that are not learned by remote control. It is as if our understanding of our own physiology depended only on what we could see from the outside—as if the precursors to modern medicine hadn't begun to cut corpses up to see what was going on inside. This is not for everyone, this direct engagement stuff. Many would prefer their meat and vegetables highly processed and neatly packaged, ready to go from the freezer to the microwave to the table. We don't all need to spend months with cadavers to become knowledgeable about how we function. But we do need people who do know where food comes from—and with what struggles, moral and otherwise, and at what costs, moral and otherwise—just as we need people who have spent months cutting up cadavers so that when something goes wrong with us, they will know our guzzle from our zatch. These are mortal stakes.

Most people do not hunt. That is as true today as it was two hundred or two thousand years ago. But two hundred or two thousand years ago, most knew directly where their next meal was coming from. They didn't need to hunt or even to know a hunter to know this. Their connections to the seamier and less pleasant sides of life were palpable and direct. Our connections are anything but direct. In this sense, we need hunters more than our ancestors did because we need to be reminded about the manifold ways in which we are connected to and dependent upon nature, a nature that is resistant to our claims on it; a nature that is, it seems, bound to frustrate our designs, whether with floods, ferocious storms, or subtle changes in climate that alter growing seasons. Farmers know these things, but they now are fewer even than the hunters among us. Gardeners know these things, too, and luckily, there are plenty of gardeners in our midst. But gardening, for all that it teaches us about nature and about how precarious our grip on nature really is, can't bear all the weight of connecting us directly to nature. Gardening is, herbicides, insecticides, and various devices to keep wildlife away notwithstanding, a gentle form of engagement. We also need to be reminded that our life on this planet is not secured through gentleness alone. Animals—lovely, graceful, canny animals—must give their lives so that we may live.

This is implacably so, even if no human ever eats another morsel of wild flesh. Wildlife is "displaced" every time a new subdivision arises from once open land. Creatures by the untold millions die each year beneath our automobile tires. The farms from which we get our food are possible to the extent that we have stolen habitat that would otherwise be home to wildlife. Moreover, ending sport hunting would not end the deliberate killing of wildlife. Those wild animals whose habits and fecundity combine to make them nuisances or threats will still be killed. Instead of being killed by men and women who admire the animals they hunt and who derive deep satisfaction from the hunt, they will be killed by hired exterminators who are unlikely to have anything remotely approaching the regard for the wild that hunters have.* I am quite sure this distinction is lost on the animals who will be killed. It can hardly be a matter of consequence to a white-tailed deer whether she finds her woodland home converted to a shopping mall or she collides with a car or is felled by a sharpshooter's bullet or a

---

*An anecdote from my hometown, Amherst, Massachusetts, makes this point all too forcefully. Though it involves trapping, not hunting, the parallel is strong. Several years ago the voters passed a ballot measure sponsored by animal rights groups banning the use of body-gripping traps (among its other measures). The result was essentially to end the traditional trapping of beaver. The result, predictably, has been an explosion in the beaver population. If beaver become a threat to public health or safety, they can be trapped after a lengthy review process has concluded that there is no alternative. Beaver appeared in one of the town's watersheds, and concerns grew that the beaver would cause the bacteria levels in the town reservoir to rise to unacceptable levels. The town sought a permit to trap, and it was granted. Fourteen beaver were trapped by town workers, and their carcasses were unceremoniously deposited in the town landfill. How this is preferable to allowing recreational trappers to trap beaver and prepare their pelts for sale I leave to the reader's imagination. What is clear is that the beaver has gone from an honored game animal to trash.

hunter's arrow. Dead is, after all, dead. But such distinctions as these surely should matter to us.

Sport hunting not only sustains an active participation in nature, it helps to sustain and embellish our appreciation for the ways wildlife and humans are indissolubly linked. We are not zoo-keepers after all, even if our protective impulses are strong. We are, as Aldo Leopold so convincingly insisted nearly fifty years ago, members of a biotic community that is based upon a volatile mix of competition and cooperation. Too much of either can create havoc. We impose mortality on wildlife, more often than not heedlessly. The virtue of sport hunting is that it is anything but heedless. Because it is deliberate, hunters can be encouraged to be discriminant and thoughtful in their pursuit of game. As much as any group in our society, hunters have helped to elevate the stature of and regard for wildlife in large measure because they see themselves as active participants, locked in competition with their quarry, while at the same time promoting the welfare of the species. This is, from the point of view of hunters, an honest relationship with wildlife. Deer should not be treasured as living lawn ornaments in suburban landscapes any more than geese should be remembered for having rendered parks and golf courses unusable by humans.

Hunting, like gardening, introduces honesty into our relationship with nature. While it leaves plenty of room for romantic idealizations, it nonetheless requires that we confront head on the ways in which we live in tension with nature and natural forces. We forget this at our own peril; as importantly, if we forget this, we also put wildlife at risk. Hunted animals are wary animals that will try to avoid humans. Animals that have not been hunted lose their fear of humans and begin showing up in places where they really do not belong, and were they hunted, they would likely avoid. The same thing can be said about backyard bird feeders, though no issue of hunting is involved. Attracting songbirds to our yards has become a major spectator "sport." We spend several billion dollars each year on feeders, seeds, and suet in order to enjoy the colorful feathers (and the antics of squirrels). The people who feed birds no doubt think they are doing the birds a favor; the enjoyment derived from watching the birds is simply a bonus. But biologists have long known that feeding puts the birds at elevated risk of predation and disease and generally lowers the fitness of the "kept" creatures. Though pure of heart, this sort of relationship to wildlife turns out to be as self-serving as hunting but, unlike hunting, it is deceptive, not honest, self-service. Because spectatorship falsifies our relationship to wildlife, it clouds our understanding of our obligations and responsibilities to the wild creatures around us. Sport hunting holds us to a higher standard.

Modern sport hunting has been crucial to the fate of wildlife, and it has helped more abstractly to keep our relationship to the wild honest. The motives and character of hunters themselves may not always rise to the high standard of honesty that is embodied in the ethic of fair chase, and as I have shown there are many social forces at work to weaken the bite of this ethical standard. Nevertheless, warts and all, the men and women who hunt remain a living repository of some very ancient truths. In a world increasingly enamored with representations and simulations of reality, sport hunters quietly remind us that there still is such a thing as authenticity, and at its core, as regards our relationship to nature, it consists of our sober recognition of the mortal stakes on which our own lives are predicated.

APPENDIX:
A COMPARISON OF THE ATTITUDES
AND BEHAVIOR OF HUNTERS AND
NONHUNTERS

All the data reported here is derived from the bi-annual survey conducted by the National Opinion Research Center (NORC), based at the University of Chicago. The NORC has been surveying random samples of Americans for decades on a wide range of topics and is among a small handful of polling organizations that caters to academic, as opposed to commercial or political, interests. The NORC's General Social Survey is widely regarded as one of the most reliable assessments of public opinion. Almost all of the data presented here derives from the General Social Survey conducted in 2000. Although 2,817 adults were interviewed, only 1,856 were asked whether or not they hunted. The data reported here are based on the responses of these 1,856 individuals, 241 (13%) of whom identified themselves as hunters. In a few instances, I have had to include data collected in earlier surveys because the questions of interest were not asked in the 2000 survey.

Following the narrative in chapter 2, the data here are gathered under four rubrics: demographic profile; attitudes toward the environment; politics and ideology; and "private life." Statistically significant differences between hunters and nonhunters are noted in brackets.

DEMOGRAPHIC PROFILE

	Hunters (%)	Nonhunters (%)
Male	81	38
White	95	77
Marital status		
Married	48	45
Divorced/widowed	26	31
Never married	25	25
Age		
<25	13	9
25–35	27	21
36–55	38	41
56–65	15	10
66 and over	8	18
Education		
<High school	13	16
High school	61	53
Some college	8	7
College graduate	12	16
Graduate/professional degree	5	8
Occupation		
White collar	50	62
Blue collar	34	20
Farm/forest	3	2
Service	13	16
Income		
<$24,999	25	40
$25,000–49,999	40	28
$50,000 and up	35	32
Residence		
Big city	8	20
Small city	51	42
Suburbs	17	24
Village/rural	23	14
Region		
East	12	21
Midwest	33	22
South	35	36
West	20	21
Religion		
Protestant	58	54
Catholic	21	25
Jewish	0.4	3
Other	4	5
None	16	13

ATTITUDES TOWARD THE ENVIRONMENT

	Hunters		Nonhunters	
	M (%)	F (%)	M (%)	F (%)
How willing would you be to pay much higher prices in order to protect the environment?				
Willing	48	27	48	39
Can't choose	30	53	31	33
Not willing	22	20	21	28
Almost everything we do in modern life harms the environment.				
Agree	46	53	47	44
Not sure	12	33	22	26
Disagree	42	13	32	30
Modern science will solve our environmental problems with little change to our way of life.				
Agree	30	7	28	18
Not sure	37	60	26	35
Disagree	33	33	46	48
Economic growth always harms the environment.				
Agree	21	20	18	20
Not sure	28	53	32	35
Disagree	51	27	49	45
For environmental problems there should be international agreements that America and other countries should be made to follow.				
Agree	76	73	79	72
Not sure	11	27	17	22
Disagree	13	—	4	6

[Difference between males is statistically significant.]

	Hunters		Nonhunters	
Some countries are doing more to protect the world than other countries are. In general, do you think that America is doing more than enough, about right, or too little?				
More than enough	17	14	15	8
About right	41	29	37	41
Too little	37	43	42	41
Can't choose	5	14	6	10

Select: (1) Government should let ordinary people decide for themselves how to protect the environment even if that means they don't always do the right thing; or (2) Government should pass laws to make ordinary people protect the environment,

(continued)

	Hunters		Nonhunters	
	M (%)	F (%)	M (%)	F (%)
even if it interferes with peoples' right to make their own decisions.				
People decide	32	21	27	23
Government laws	41	29	49	50
Can't choose	24	50	24	27
Select: (1) Government should let businesses decide for themselves how to protect the environment even if that means they don't always do the right thing; or (2) government should pass laws to make business protect the environment, even if it interferes with businesses' right to make their own decisions.				
Business decide	8	7	10	9
Government laws	76	64	74	74
Can't choose	16	29	16	17
In the last five years have you given money to an environmental group?				
Yes	17	21	23	22
No	83	79	77	78
Select: (1) Government should let ordinary people decide for themselves how to protect the environment even if that means they don't always do the right thing; or (2) Government should pass laws to make ordinary people protect the environment, even if it interferes with peoples' right to make their own decisions.				
People decide	32	21	27	23
Government laws	41	29	49	50
Can't choose	24	50	24	27
Select: (1) Government should let businesses decide for themselves how to protect the environment even if that means they don't always do the right thing; or (2) government should pass laws to make business protect the environment, even if it interferes with businesses' right to make their own decisions.				
Business decide	8	7	10	9
Government laws	76	64	74	74
Can't choose	16	29	16	17
In the last five years have you given money to an environmental group?				
Yes	17	21	23	22
No	83	79	77	78

POLITICS AND IDEOLOGY

	Hunters M (%)	Hunters F (%)	Nonhunters M (%)	Nonhunters F (%)
Party identification				
Strong Democrat	7	13	15	17
Democrat	21	16	29	32
Independent	16	31	21	21
Republican	32	33	26	23
Strong Republican	24	7	8	9

[Differences between both males and females are statistically significant.]

	Hunters M	Hunters F	Nonhunters M	Nonhunters F
Vote in 1996				
Clinton	36	37	53	64
Dole	46	33	33	25
Perot	17	30	11	10
Didn't vote	1	—	2	1

[Differences between both males and females are statistically significant.]

Where do you place yourself on the scale from liberal to conservative?

	Hunters M	Hunters F	Nonhunters M	Nonhunters F
Liberal	16	21	29	27
Moderate	40	42	37	43
Conservative	45	37	34	30

[Difference between males is statistically significant.]

Do you favor or oppose the death penalty for persons convicted of murder?

	Hunters M	Hunters F	Nonhunters M	Nonhunters F
Favor	85	70	66	57
Oppose	10	30	27	33
Undecided	5	—	7	10

[Differences between both males and females are statistically significant.]

Would you favor or oppose a law which would require a person to obtain a police permit before he or she could buy a gun?

	Hunters M	Hunters F	Nonhunters M	Nonhunters F
Favor	50	72	78	87
Oppose	49	28	20	10
Don't know	1	—	2	3

[Differences between both males and females are statistically significant.]

Abortion should be available if a woman wants it for any reason.

	Hunters M	Hunters F	Nonhunters M	Nonhunters F
Yes	28	33	41	38
No	68	63	54	57
Undecided	4	4	6	4

[Difference between males is statistically significant.]

Do you agree or disagree that methods of
birth control should be available to teenagers
between 14 and 16 if their parents do not approve?

(continued)

	Hunters		Nonhunters	
	M (%)	F (%)	M (%)	F (%)
Agree	54	61	62	58
Disagree	44	30	36	40
Undecided	1	9	2	2

Do you agree or disagree with this statement: Most men are better suited emotionally for politics than most women.

Agree	26	22	20	22
Disagree	64	78	71	70
Not sure	10	—	8	7

On the average, blacks have worse jobs, income, and housing than whites. Do you think these differences are because blacks have less inborn ability to learn?

Yes	16	9	13	13
No	82	91	84	83
Don't know	2	—	4	4

Do you agree, neither agree nor disagree, or disagree with the following statement: Irish, Italians, Jewish and many other minorities overcame prejudice and worked their way up. Blacks should do the same without special favors. (1998)

Agree	81	61	67	70
Disagree	7	25	15	15
Neither	12	14	18	16

[Difference between males is statistically significant.]

How would you respond to a close relative marrying a black person?

Accept	26	32	29	32
Oppose	34	32	30	31
Stay neutral	40	37	41	37

What about sexual relations between two adults of the same sex—do you think it is always wrong, sometimes wrong, or not wrong at all?

Always wrong	72	52	57	56
Sometimes wrong	5	4	8	8
Not wrong	18	30	28	28
Don't know	5	13	8	9

[Difference between males is statistically significant.]

How much confidence do you have in the Federal Government?

A great deal	8	17	17	16

(continued)

	Hunters		Nonhunters	
	M (%)	F (%)	M (%)	F (%)
Only some	43	48	51	51
Hardly any	48	35	30	28
Don't know	2	—	2	5

Do you consider the amount of federal taxes you have to pay as too high, about right, or too low?

	Hunters		Nonhunters	
Too high	74	76	61	63
About right	25	24	34	31
Too low	1	—	2	1
Don't know	—	—	4	5

[Difference between males is statistically significant.]

Government should do something to reduce income differences between rich and poor.

	Hunters		Nonhunters	
Should	36	48	36	45
Should not	44	35	44	31
Undecided	20	17	20	23

Do you think that the government should improve living standards or that people should take care of themselves?

	Hunters		Nonhunters	
Government should	26	35	23	30
Self-help	39	2	28	25
Can't decide	35	39	49	46

PRIVATE LIFE

	Hunters		Nonhunters	
	M (%)	F (%)	M (%)	F (%)

Do you regard yourself a religious fundamentalist, moderate, or liberal?

	Hunters		Nonhunters	
Fundamentalist	30	28	24	34
Moderate	40	44	40	39
Liberal	30	28	36	27

About how often do you pray?

	Hunters		Nonhunters	
At least daily	41	64	43	65
Several times a week	29	29	26	19
Less than weekly	30	6	31	15
Never	—	—	1	1

How often do you attend religious services?

	Hunters		Nonhunters	
Never	20	27	26	19

(continued)

	Hunters		Nonhunters	
	M (%)	F (%)	M (%)	F (%)
<once a year	10	2	9	7
1–2 times a year	30	36	26	23
At least once a month	20	18	17	23
Weekly	20	18	22	28

Where would you place your image of the world on the following scale: The world is basically filled with evil and sin; or there is much goodness in the world which hints at God's goodness.

Evil	14	9	15	12
Goodness	48	54	58	60
Can't decide	38	36	27	28

Have you seen an X-rated movie in the last year?

Yes	30	30	34	17

In the past thirty days, how often have you visited a website for sexually explicit materials?

Never	76	100	79	96
Once or twice	20	—	13	2
Three or more times	5	—	8	2

Thinking about the time since your 18th birthday, have you ever had sex with a person you paid or who paid you for sex?

Yes	14	—	19	2

Have you ever had sex with someone other than your husband or wife while you were married?

Yes	18	10	18	10
No	53	66	52	68
Not wed	29	24	31	22

About how often did you have sex during the past twelve months?

Not at all	14	8	18	28
Once or twice	18	10	18	18
1–3 times a month	17	12	18	14
At least once a week	44	52	37	35
More than once a week	6	18	9	5

[Difference between females is statistically significant.]

Now think about the past five years— how many sex partners have you had?

0	5	2	11	18
1	54	57	50	53
2–3	18	26	17	21
More than 3	23	14	22	8

[Difference between females is statistically significant.]

(continued)

	Hunters M (%)	Hunters F (%)	Nonhunters M (%)	Nonhunters F (%)
Have your sex partners in the last five years been:				
Male	2	92	5	96
Both male & female	—	5	2	2
Female	98	2	93	2
[Difference between males is statistically significant.]				
Taken all together, how would you say things are these days—would you say you are very happy, pretty happy, or not too happy?				
Very happy	32	49	31	31
Pretty happy	60	48	57	59
Not too happy	8	4	12	11
In general, do you find life exciting, pretty routine, or dull?				
Exciting	58	54	49	42
Pretty routine	38	44	46	53
Dull	4	2	4	6
Do you even have occasion to use any alcoholic beverages such as liquor, wine, or beer, or are you a total abstainer? (1994)				
Drink	89	60	74	64
Abstain	11	40	26	36
Do you sometimes drink more than you think you should? (1994)				
Yes	58	33	45	30
No	42	67	55	70
Would you say your own health, in general, is excellent, good, fair, or poor? (1996)				
Excellent	35	39	32	30
Good	45	37	45	48
Fair	16	18	18	17
Poor	4	6	5	4
In the past seven days, how often have you felt that you couldn't shake the blues? (1996)				
0	54	50	54	46
1–2 days	29	—	28	27
3 or more days	17	50	19	27
In the past seven days, how often have you felt really angry or irritated at someone? (1996)				
0	36	25	34	35
1–2 days	34	39	46	45
3 or more days	30	36	20	20
[Difference between males is statistically significant.]				

(continued)

	Hunters M (%)	Hunters F (%)	Nonhunters M (%)	Nonhunters F (%)
Did the situation that made you angry involve your spouse, boyfriend/girlfriend, or partner? (1996)				
Yes	4	27	8	19
After you got angry, did you have a drink or take a pill? (1996)				
Yes	4	—	10	5
After you got angry, did you yell or hit something to let out pent-up feelings? (1996)				
Yes	9	14	8	8

WORKS CITED

Adair, Holiday E. 1995. "The Correlation between Hunting and Crime: A Comment." *Society and Animals* 3:189–95.

Albanese, Catherine. 1990. *Nature Religion in America*. Chicago: University of Chicago Press.

Baker, Ron. 1985. *The American Hunting Myth*. New York: Vantage Press.

Bilger, Burkhard. 2001. "A Shot in the Ark." *New Yorker,* March 25, 74–83.

Bergman, Charles. 1996. *Orion's Legacy: A Cultural History of Man as Hunter.* New York: Dutton.

Blüchel, Kurt G. 1997. *Game and Hunting*. Cologne, Germany: Könemann.

Blumstein, Alfred, and Joel Wallman, eds. 2000. *The Crime Drop in America*. New York: Cambridge University Press.

Botkin, Daniel. 1990. *Discordant Harmonies: A New Ecology for the Twenty-first Century*. New York: Oxford University Press.

Brandt, Anthony. 1997. "Not in My Backyard." *Audubon*, September–October, 58–62, 86–87, 102–3.

Brody, Hugh. 1982. *Maps and Dreams*. New York: Pantheon.

———. 2000. *The Other Side of Eden: Hunters, Farmers, and the Shaping of the World*. New York: North Point Press.

Brown, Beverly A. 1995. *In Timber Country*. Philadelphia, Pa.: Temple University Press.

Campbell, Joseph. 1988. *Historical Atlas of World Mythology*. Vol. 1, pt. 2, *Mythologies of the Great Hunt*. New York: Harper & Row.

Cartmill, Matt. 1993. *A View to a Death in the Morning: Hunting and Nature through History*. Cambridge: Harvard University Press.

Clifton, Merritt. 1994. "Hunters and Molesters." *Animal People* 3:1, 7–9.

Church, Peggy Pond. 1960. *House at Otowi Bridge*. Albuquerque: University of New Mexico Press.

Cronon, William. 1983. *Changes in the Land*. New York: Hill and Wang.

Dizard, Jan E. 1999. *Going Wild: Hunting, Animal Rights, and the Contested Meaning of Nature*. Amherst: University of Massachusetts Press.

Dizard, Jan E., and Robert M. Muth. 2001. "The Value of Hunting: Connections to a Receding Past and Why These Connections Matter." *Transactions of the Sixty-sixth North American Wildlife and Natural Resources Conference* 66:154–70.

Duda, Mark Damian, Steven J. Bissell, and Kira C. Young. 1998. *Wildlife and the American Mind*. Harrisonburg, Va.: Responsive Management.

Ehrenreich, Barbara. 1997. *Blood Rites: Origins and History of the Passions of War*. New York: Metropolitan Books.

Ezekiel, Raphael S. 1995. *The Racist Mind: Portraits of American Neo-Nazis and Klansmen*. New York: Viking.

Glasser, Barry. 1999. *The Culture of Fear: Why Americans Are Afraid of the Wrong Things*. New York: Basic Books.

Gray, Gary G. 1993. *Wildlife and People: The Human Dimensions of Wildlife Ecology*. Urbana: University of Illinois Press.

Guttmann, Allen. 1978. *From Ritual to Record: The Nature of Modern Sports*. New York: Columbia University Press.

———. 1986. *Sports Spectators*. New York: Columbia University Press.

Heberlein, Thomas A., and E. Thomson. 1996. "Changes in U.S. Hunting Participation, 1980–1990." *Transactions of the Twenty-second Congress of the International Union of Game Biologists*. 00:373–77.

Henslin, James N., and Mae A. Biggs. 1995. "Behavior in Public Places: The Sociology of the Vaginal Examination." In *Down to Earth Sociology*, 8th ed., edited by James N. Henslin, 201–12. New York: Free Press.

Herman, Daniel Justin. 2001. *Hunting and the American Imagination*. Washington, D.C.: Smithsonian Institution Press.

Izaak Walton League. 1999. *Hunting Ethics Land-Access Project: A Report by the Izaak Walton League of America and Responsive Management*. Gaithersburg, Md.: Izaak Walton League of America.

Jackson, Bob, and Robert Norton. 1987. "Hunting as a Social Experience." *Deer and Deer Hunting* 11:38–51.

Kellert, Stephen. 1978. "Attitudes and Characteristics of Hunters and Anti-Hunters." *Transactions of the Forty-third North American Wildlife and Natural Resources Conference* 43:412–23.

———. 1996. *The Value of Life: Biodiversity and Human Society*. Washington, D.C.: Island Press.

Kempton, Willett, James S. Boster, and Jennifer A. Hartley. 1995. *Environmental Values in American Culture*. Cambridge: MIT Press.

Kerasote, Ted. 1993. *Bloodties: Nature, Culture, and the Hunt*. New York: Random House.

————. 1997. *Heart of Home*. New York: Villard.

Kheel, Marti. 1995. "License to Kill: An Ecofeminist Critique of Hunters' Discourse." In *Animals and Women*, edited by C. J. Adams and J. Donovan, 85–125. Durham, N.C.: Duke University Press.

Krech, Shepard, III. 1999. *The Ecological Indian: Myth and History*. New York: Norton.

Langston, Nancy. 1995. *Forest Dreams, Forest Nightmares: The Paradox of Old Growth in the Inland West*. Seattle: University of Washington Press.

Lefkowitz, Bernard. 1997. *Our Guys*. New York: Vintage Books.

Lembcke, Jerry. 1998. *The Spitting Image: Myth, Memory, and the Legacy of Vietnam*. New York: New York University Press.

Leopold, Aldo. 1949. *A Sand County Almanac*. New York: Oxford University Press.

Lutz, Ralph H. 1990. *The Nature Fakers: Wildlife, Science and Sentiment*. Golden, Colo.: Fulcrum.

Martin, Calvin. 1978. *Keepers of the Game: Indian-Animal Relationships and the Fur Trade*. Berkeley and Los Angeles: University of California Press.

Miller, John M. 1992. *Deer Camp: Last Light in the Northeast Kingdom*. Cambridge: MIT Press.

Minnis, Donna L. 1997. "The Opposition to Hunting: A Typology of Beliefs." *Transactions of the North American Wildlife and Nature Resources Conference* 62:346–60.

Mitchell, John. 1980. *The Hunt*. New York: Knopf.

Muth, Robert M., and Wesley V. Jamison. 2000. "On the Destiny of Deer Camps and Duck Blinds: The Rise of the Animal Rights Movement and the Future of Wildlife Conservation." *Wildlife Society Bulletin* 28:841–51.

Nash, Roderick. 1982. *Wilderness and the American Mind*. New Haven: Yale University Press.

Nelson, Richard. 1990. *The Island Within*. Washington, D.C.: Island Press.

————. 1997. *Heart and Blood: Living with Deer in North America*. New York: Knopf.

Organ, John F., et al. 1998. "Fair Chase and Humane Treatment: Balancing the Ethics of Hunting and Trapping." *Transactions of the Sixty-third North American Wildlife and Natural Resources Conference* 63:528–43.

Peterson, David. 2000. *Heartsblood: Hunting, Spirituality, and Wildness in America*. Washington, D.C.: Island Press.

Posewitz, Jim. 1994. *Beyond Fair Chase*. Helena, Mont.: Falcon Press.

————. 1999. *Inherit the Hunt: A Journey into the Heart of American Hunting*. Helena, Mont.: Falcon Press.

Poten, Constance J. 1991. "A Shameful Harvest: America's Illegal Wildlife Trade." *National Geographic* 108:106–32.

Reiger, George. 1999. "Snow Goose Wars." *Field and Stream,* June: 28, 30.

Reiger, George. 2000. "What Good Is Hunting?" *Field and Stream,* December: 26–29.

Reiger, John F. 1975. *American Sportsmen and the Origins of Conservation.* New York: Winchester Press.

Russell, Sherman Apt. 1993. *Kill the Cowboy: A Battle of Mythology in the New West.* Reading, Mass.: Addison-Wesley.

Schmitt, Peter J. 1969. *Back to Nature: The Arcadian Myth in Urban America.* Baltimore: Johns Hopkins University Press.

Schneider, Paul. 1997. *The Adirondacks: A History of America's First Wilderness.* New York: Henry Holt.

Shepard, Paul. 1973. *The Tender Carnivore and the Sacred Game.* Athens: University of Georgia Press.

Stange, Mary Zeiss. 1997. *Woman the Hunter.* Boston: Beacon Press.

Stange, Mary Zeiss, and Carol K. Oyster. 2000. *Gun Women: Firearms and Feminism in Contemporary America.* New York: New York University Press.

Swan, James A. 1995. *In Defense of Hunting.* New York: Harper Collins.

Thoreau, Henry David. 1992. *Walden and Other Writings of Henry David Thoreau.* Edited by Brooks Atkinson. New York: Modern Library.

Tober, James. 1981. *Who Owns the Wildlife? The Political Economy of Conservation in Nineteenth-Century America.* Westport, Conn.: Greenwood.

Trefethen, James. 1975. *An American Crusade for Wildlife.* Alexandria, Va.: Boone and Crockett Club.

Turner, Patricia. 1993. *I Heard It through the Grapevine: Rumor in African-American Culture.* Berkeley and Los Angeles: University of California Press.

Vialles, Noelie. 1994. *Animal to Edible.* Translated by J. A. Underwood. New York: Cambridge University Press.

Wagner, Hans-Peter. 1982. *Puritan Attitudes toward Recreation in Early Seventeenth-Century New England.* Frankfurt am Mein: P. Lang.

Warren, L. E. 1997. *The Hunter's Game: Poachers and Conservationists in Twentieth-Century America.* New Haven: Yale University Press.

Worster, Donald. 1977. *Nature's Economy: A History of Ecological Ideas.* New York: Cambridge University Press.

Williams, Joy. 1990. "The Killing Game." *Esquire,* October, 113–28.

Zuckerman, Michael. 1977. "Pilgrims in the Wilderness: Community, Modernity, and the Maypole at Merry Mount." *New England Quarterly* 1:255–75.

INDEX